INTELLIGENT TRANSPORTATION SYSTEMS
From Good Practices to Standards

To the women of my life

wife and daughters

INTELLIGENT TRANSPORTATION SYSTEMS
From Good Practices to Standards

Editor

Paolo Pagano

Consorzio Nazionale Interuniversitario per le
Telecomunicazioni (CNIT)
Director of CNIT Laboratory at the Livorno Port Authority,
Livorno (Italy)

CRC Press
Taylor & Francis Group
Boca Raton London New York

CRC Press is an imprint of the
Taylor & Francis Group, an **informa** business

A SCIENCE PUBLISHERS BOOK

CRC Press
Taylor & Francis Group
6000 Broken Sound Parkway NW, Suite 300
Boca Raton, FL 33487-2742

First issued in paperback 2020

© 2017 by Taylor & Francis Group, LLC
CRC Press is an imprint of Taylor & Francis Group, an Informa business

ISBN-13: 978-1-4987-2186-8 (hbk)
ISBN-13: 978-0-367-78282-5 (pbk)

Library of Congress Cataloging-in-Publication Data

Names: Pagano, Paolo, editor.
Title: Intelligent transportation systems / editor, Paolo Pagano.
Description: Boca Raton : CRC Press, 2016. | Includes bibliographical references and index.
Identifiers: LCCN 2016016904 | ISBN 9781498721868 (hardcover : alk. paper)
Subjects: LCSH: Transportation--Europe--Planning. | Intelligent transportation systems--Europe.
Classification: LCC HE242.A2 .I57 2016 | DDC 388.3/12094--dc21
LC record available at https://lccn.loc.gov/2016016904

Visit the Taylor & Francis Web site at
http://www.taylorandfrancis.com

and the CRC Press Web site at
http://www.crcpress.com

Preface

In the Horizon 2020 programme being implemented by the European Commission, a major social challenge is that of "achieving a European transport system that is resource-efficient, environmentally friendly, safe and 'seamless' for the benefit of citizens, the economy and society". To accomplish this challenge, the Commission has promoted an "Integrated effort"* including both governance and technological aspects. The governance is expected to "transcend the scope and scale of purely national efforts", in order to set up *i-Transport systems* complying with "geographical continuity, standardisation and interoperability of service". It will also be beneficial to avoid the "emergence of a patchwork of ITS applications and services", while it is expected to remove the limitations of "technological improvements involving individual vehicles or infrastructure components and sub-systems". In addition, "technological solutions must be found at the level of the interactions between the various constituents of transport systems, including users, and their optimal combination".

Intelligent Transport Systems (ITS), i.e. ICT applied to transportation and mobility, are considered the turnkey for a sustainable growth. In a nutshell, ITS's target is to improve vehicle safety, increase the efficiency of the multimodal mobility of people and freight, and demonstrate eco-sustainable mobility by limiting the environmental impact.

Implementation of ITS requires active participation by the industry to overcome the current limitations of the very fragmented solutions that are there in the market.

The FP6 and FP7 European Commission programmes have supported innovation and cooperation actions by funding a number of projects impacting the standardization process of ITS technologies ready to be integrated in operational deployments.

European experimentation has focused on Autonomous Vehicles (from ICT call 4, 2007) and ICT for Transport (from ICT call 6, 2009).

---

* European Commission C (2013) 8631 of 10 December 2013, Work Programme – Part XI – Societal Challenges, Smart, Green and Integrated Transport.

Therefore, numerous projects have been funded, notably "Co-operative Vehicle – Infrastructure Systems" (CVIS), "Geo-addressing and geo-routing for vehicular communications" (GEONET), "Cooperative systems for road safety: Smart vehicles on smart roads" (SAFESPOT), and An integrated wireless and traffic platform for real-time road traffic management solutions" (ITETRIS). On the one hand, they set up the technological foundations of Vehicle-to-Vehicle (V2V) and Vehicle-to-Infrastructure (V2I) communication, on the other hand they contributed to the development of such applications generally encompassed in the so-called Cooperative ITS (C-ITS).

The Commission has also supported collaborative projects focusing on autonomous driving, as "Highly Automated Vehicles for Intelligent Transport" (HAVEit), "Safe Road Trains for the Environment" (SARTRE), "Advancing automated driving" (ADAPTIVE), etc.

Field Operational Tests (FOTs) started in 2008 focusing both on Advanced Driving assistance Systems (ADAS) and C-ITS. Large scale projects like EUROFOT, TELEFOT, DRIVE-C2X, etc. have generated big data sets, which have been used in turn to assess the maturity of such technologies. Support Actions like FOT-NET and FOT-NET DATA are disseminating a common (V-shaped) methodology to develop ITS pilots and analytically quote the results through a set of standardized Key Performance Indicators.

Data from ITS pilots are being collected and made available to increase their utility via increasingly sophisticated analysis data comparison and correlation, and allowing collaboration across communities of experts. Software and calculus tools are being proposed to allow fast data inspection, reproducible analysis, correlation with third-party data sets, sophisticated modelling etc.

At the normative side, the European frequency band allocation (Commission Decision 2008/671/EC) was followed in October 2009 by an EC standardization mandate (M/453) to support a European Community wide implementation and deployment of Cooperative ITS. The Release 1 of these standards has been published by ETSI TC ITS and CEN TC 278 in February 2014 including standardized testing procedures widely adopted by ETSI for compliance and interoperability campaigns (known as Plugtests™)*.

A major challenge for the next decade is that of bridging the gap of ITS with the world of distributed sensing and Machine-to-Machine communications. The Alliance for Internet of Things Innovation (AIOTI)† aims at integrating into an open (standard) platform all technologies

---

* See http://www.etsi.org/about/what-we-do/plugtests. Last access on May 12th, 2016.

†See http://www.aioti.eu. Last access on May 12th, 2016.

related to sensing and actuation; applications include agriculture, wearable devices, smart manufacturing, smart cities, and smart mobility.

In the domain of mobility[*] the alliance aims at a successful deployment of safe and automated vehicles (up to SAE level 5[†]) "using IoT technology in different use case scenarios, using local and distributed information and intelligence". Information is gathered from "safety critical vehicle services, advanced sensors/actuators, navigation and cognitive decision-making technology, interconnectivity between vehicles (V2V) and vehicle to infrastructure (V2I) communication".

An effective integration of these technologies will be validated through Large Scale Pilots (as from the Horizon 2020 Work Programme 2016-17)[‡] appointed to demonstrate dependability, robustness and resilience over a long period of time and under a large variety of conditions.

If technologies are integrated into accessible, well-document, open, and standard platforms, it will be possible to implement large scale ITS in strategic environments like metropolises, airports, seaports, etc.

If relevant logistics and transport nodes feature standard ITS services, more links, new trades and new wealth are expected to show up, fulfilling the expectations sometimes summarized in the "Mobility for Growth" slogan.

March, 2016   **Paolo Pagano**

---

[*] See the WG9 report: https://ec.europa.eu/digital-agenda/en/news/aioti-recommendations-future-collaborative-work-context-internet-things-focus-area-horizon-2020. Last access on May 12th, 2016.

[†] SAE J301614, http://standards.sae.org/j3016_201401. Last access on May 12th, 2016

[‡] European Commission Decision C (2015) 6776 of 13 October 2015, Work Programme - Part XVII - Cross-cutting activities (Focus Areas)

Contents

Development of an ITS-G5 Station, from the Physical to the MAC Layer

João Almeida[1,2*], Joaquim Ferreira[1,3] and Arnaldo S.R. Oliveira[1,2]

[1]Instituto de Telecomunicações, Campus Universitário de Santiago
3810-193 Aveiro, Portugal
[2]DETI – University of Aveiro, Campus Universitário de Santiago
3810-193 Aveiro, Portugal
[3]ESTGA – University of Aveiro, 3754-909 Águeda, Portugal

Abstract

Wireless vehicular networks for cooperative Intelligent Transportation Systems (ITS) have raised widespread interest in the last few years, due to their potential applications and services. Cooperative applications with data sensing, processing and communication provide an unprecedented potential to improve vehicle and road safety, passenger's comfort and efficiency of traffic management and road monitoring. Safety, efficiency and comfort applications envisaged for ITS exhibit tight latency and throughput requirements. For example safety critical services require guaranteed maximum latencies lower than 100 ms, while most infotainment applications require QoS support and data rates higher than 1 Mbit/s. In this context, BRISA, a motorway operator, challenged a team from Institute of Telecommunications (IT) to contribute with research in this area, to work specifically on the then emergent IEEE 802.11p amendment. Back then, standards in this field were (and still are) evolving and only a very limited number of Commercially Off The Shelf (COTS) components were available. Available COTS chip sets implemented an incomplete stack and, more importantly, have closed implementations (black box) of the standard, with a limited access to the API and programming model. This was an impairment to BRISA track record on open access to technology and independence from manufacturers. On the other hand, from the point

*Corresponding author: jmpa@ua.pt

of view of a research institution as IT, having the chance of developing from scratch, new technology and draft protocol implementations was considered quite relevant, as it potentiates the inclusion of innovative solutions beyond standards. Examples of such innovative solutions include support for real-time operation and fault-tolerance mechanisms. That was the rationale behind the development of a new ITS-G5 station, which will be described in a tutorial way in this chapter. The chapter presents the architecture of the PHY and MAC layers developed, including the RF front-end, the baseband PHY processing chains, the time sensitive lower MAC implementation and the software upper MAC. The most important design decisions, the validation methodologies and the interoperability tests are also discussed. The chapter ends with the presentation of the real-time extensions and primitives of the platform enabling the implementation of real-time protocols.

1. Motivation

Road traffic accidents are a serious problem worldwide, being one of the major causes of death. During 2014, there were about 26,000 fatalities, only inside the European Union borders [1]. Vehicular communication systems are being developed with the intention of improving traffic safety, by extending the sensing range of the vehicles via wireless transmission of information. Vehicular communications are an important field of research in the area of Intelligent Transportation Systems (ITS). Future ITS will require dependable wireless communications among vehicles and between vehicles and the roadside infrastructure. Vehicular communication systems can be more effective in preventing road accidents than the case where vehicles act individually to achieve the same goal. This is due to the cooperative techniques that can be exploited when vehicles and the roadside stations are aware of the situation of other parties (e.g. location, speed and heading). As an example of this class of safety applications, chain collisions could be avoided if the information about the first crash is disseminated by all the other nodes in the vicinity of the accident. Dedicated Short-Range Communications (DSRC) is a wireless technology that has been designed to support such visionary applications based on vehicle-to-vehicle (V2V) and vehicle-to-infrastructure (V2I) communications. Vehicular communications supported by DSRC systems operate in the 5.9 GHz reserved spectrum band and exhibit an approximate maximum range of 1000 m.

Safety critical vehicular applications are inherently hard real-time systems for which the failure of meeting a deadline can lead to severe damage and pose significant threat to human life [2]. The first few steps in attempting to provide the necessary quality of service were given by acknowledging that vehicular communications require specifically designed communication systems in order to cope with high network dynamics and short connection times [3]. Some standards attempting to solve these issues were developed, namely the IEEE WAVE and ETSI ITS-G5 [4], both based on the same IEEE 802.11 physical layer [5] but containing specific features tailored for vehicular environment use, such as: reduced bandwidth channels, increased maximum power and segregated frequency bands. The main differences between these two standards are the medium access mechanisms, with WAVE using only the already existing Carrier Sense Multiple Access with Collision Avoidance (CSMA/CA) and Enhanced Distributed Channel Access (EDCA) mechanisms, while ITS-G5 adding Decentralized Congestion Control (DCC) on top of them, as an upper layer function that strive to limit the channel load by changing MAC and other transmission parameters. However, these methods have been proved insufficient in dealing with what is regarded as one of the most challenging issues in providing hard real-time characteristics to vehicular communication systems: deterministic channel access under high load conditions [3, 6, 7]. In this context, ITS stations need to be equipped with mechanisms to support deterministic communications at the lowest possible layer of the communications stack, in order to provide bounded transmission delays, reduced jitter and strict traffic isolation between safety and infotainment message streams, i.e., support for deterministic Medium Access Control (MAC) protocols.

Despite the use of dedicated spectrum and specially designed devices, the current standards governing VANET still fall short of providing dependable MAC, as desirable if this technology is to be used in support of safety applications. Moreover, study of novel MAC mechanisms is hampered by the fact that the highly dynamic channel makes it difficult to conduct simulated experiments properly [3], calling for the use of actual devices that implement the relevant standards but at the same time allow experimentation with new mechanisms, beyond current standards. Despite the growing popularity of Software Defined Radio (SDR), open implementations of IEEE 802.11p, capable of supporting the operation of deterministic MAC protocols both at hardware and software level, are not widely available. Furthermore, existing implementations of IEEE 802.11p lack support for deterministic Medium Access Control (MAC) protocols that require a set of requirements, which, to the best of one's knowledge, are not usually available in COTS transceivers.

The work presented in this chapter describes some contributions made within the scope of two projects: HEADWAY (Highway Environment ADvanced WArning sYstem) and ICSI (Intelligent Cooperative Sensing for Improved traffic efficiency). The former project aims to implement a communications platform compliant with IEEE 802.11p standard and the IEEE 1609.x sub-standards, while the latter is focused on the development of a deterministic MAC protocol, timely security services and fault-tolerance mechanisms for cooperative systems based on IEEE WAVE and ETSI ITS-G5. The main contribution of this chapter is a detailed description of the design of a ITS-G5 station, which was built from scratch, with the development of all the OSI layers, from the Physical (PHY) up to the Application. For this purpose, a SDR approach was adopted where the time-critical operations are implemented in hardware (Physical and Lower MAC layers) and the other operations are implemented in software (Upper MAC layer and above). The developed platform can operate either as a road-side unit (RSU), when integrated in the road infrastructure, or as an on-board unit (OBU), when it's used as a vehicle communications device.

Several other projects have decided to adopt a different approach to the development of IEEE WAVE/ETSI ITS-G5 compliant platforms, by extending or tweaking the functionality on top of equipment compliant with the IEEE 802.11a/g family, so that it provides the same services and functionality as the WAVE standards. That approach has obvious advantages in terms of development time, since the bulk of the work is done by a third party implementation, but the core of the device is still a black box to which the developers have no access. Such a system is therefore monolithic and hard to adopt when functional modifications, beyond existing standards, are required.

Building an IEEE WAVE/ETSI ITS-G5 station from scratch implies a much higher workload, yet it provides developers with a deep knowledge of the core functioning of the device, which in turns allows great flexibility when any kind of change or expansion is required. More importantly, the white box access to an ETSI ITS-G5 station, as opposed to the black box access provided by existing commercial off the shelf components (COTS), allows the addition of new functionalities beyond the standard. In this way it will be possible, for example, to comply fully with the standard in the Service Channel and have an enhanced operational mode in the Control Channel. This enhanced operation will rely on the deterministic behaviour provided by a real-time MAC protocol, which will require extended functionalities such as real-time Tx and Rx buffers, time-triggered message transmission without sensing the wireless medium, violating the standard inter-frame space (IFS) values, etc.

1.1 Relevant standards

There are two main protocol architectures for vehicular communication systems, one developed by the Institute of Electrical and Electronics Engineers (IEEE) and the other one from the European Telecommunications Standard Institute (ETSI), both illustrated in Fig. 1.

The protocol stack in America is denominated as IEEE Wireless Access in Vehicular Environments (WAVE), while the one in Europe is referred to as ETSI ITS-G5. Both of them rely on IEEE 802.11p, from the IEEE 802.11 family of Wi-Fi standards, for the implementation of their physical (PHY)

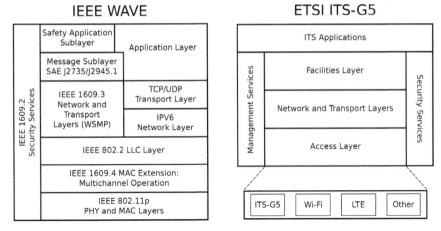

Figure 1: IEEE WAVE and ETSI ITS-G5 protocol stacks.

and medium access control (MAC) layers [5]. The physical layer is almost identical to IEEE 802.11a, using also OFDM with BPSK, QPSK, 16-QAM and 64-QAM modulations, but with double timing parameters to achieve less interference due to the multi-path propagation and the Doppler shift effect. With double timing parameters, the channel bandwidth is 10MHz instead of 20MHz, and the data rate is half, i.e., 3...27 Mbit/s instead of 6...54 Mbit/s.

The medium access control (MAC) layer adopts a carrier sense multiple access with collision avoidance (CSMA/CA), as IEEE 802.11a, but it is adjusted for the vehicular communication environments, which differ significantly from the sparse and low-velocity characteristics of a traditional Wi-Fi deployment. In vehicular environments, nodes present high mobility, some areas are often densely populated and there is non-line-of-sight frequently. As a consequence, some tweaks were introduced in the standard to allow low overhead operations, in order to guarantee fast and reliable exchange of safety messages. For example, non-IP messages that operate outside the context of a Basic Service Set (BSS) were defined, enabling quick transmission of packets by avoiding the registration and authentication procedures, commonly present in typical wireless local area networks.

The Federal Communications Commission (FCC) in the United States and the European Conference of Postal and Telecommunications Administrations (CEPT) allocated a dedicated spectrum band at 5.9 GHz (Fig. 2). In America, a bandwidth of 75 MHz was reserved, while in Europe only 50 MHz were assigned. This spectrum was divided into smaller 10 MHz wide channels and in the American case, a 5 MHz guard band at the low end was also included. As a result, there are 7 different channels for IEEE WAVE operation and 5 for the case of ETSI ITS-G5.

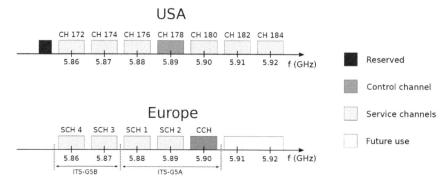

Figure 2: Spectrum allocation for vehicular communications in USA and Europe.

In Europe, 30 MHz (3 channels) are reserved for road safety in the ITS-G5A band and 20 MHz are assigned for general purpose ITS services in the ITS-G5B band. As a general rule, a control channel (CCH 178 in the USA and CCH 180 in Europe) is used exclusively for cooperative road safety and control information. The remaining channels are designated as service channels (SCH). In the United States, concerns about the reduced capacity for road safety messages led to the decision to allocate SCH 172 specifically for applications regarding public safety of life and property.

1.2 Requirements

The first requirement for the design of the ITS station is being able to operate according to both IEEE WAVE and ETSI ITS-G5 specifications. Although sharing the same physical and data link layers, already covered by IEEE 802.11, the ITS-G5 standard adds features for decentralized congestion control (DCC) to adjust the network load and avoid unstable behaviour. Also ITS-G5 requires two radios [4], one for the control channel and other for the service channel, while IEEE WAVE require just one radio [8].

In addition to the requirement of deploying a station that should be compliant with both the European and American standards, there was also a demand to develop a flexible and reconfigurable research platform able to support the implementation and validation of extensions to the standard operation of vehicular communication protocols.

For instance, detailed channel load measurements are required, in order to allow for decentralized solutions where the stations adapt their transmission patterns to the medium occupation. It is then necessary to capture detailed channel usage statistics, which should include fine grained Received Signal Strength Indication (RSSI) measurements. Another required functionality is to be able to execute run time modifications to the sensitivity level, above which the medium is regarded as busy and therefore packets are detected. This funcionality would allow the implementation of mechanisms such as DCC, where under heavy load conditions the minimum signal level for busy medium declaration is raised. This limits the effective range of the devices and reduces the coverage area and thereby the maximum amount of vehicles competing for the shared medium. This should be done during the inter-frame space, for not interfering with an ongoing message reception.

The device must also be able to disable any collision avoidance mechanisms it possesses in order to be able to transmit packets in a timely fashion and with disregard for any ongoing transmission, resembling the single shot transmission mode present in some fieldbus protocols. Although this could create forced collisions, some proposed protocols explicitly require the behaviour in order to be able to perform strict TDMA-based communication.

The communication device should also be coupled tightly to a time synchronization mechanism to support low jitter TDMA implementations. This support should be present in hardware to shield the time sensitive parts of the system from interference from the higher layers of the communication stack. That is, the transmission instant of a message should be independent from the time when it is written in the transmission buffer, to minimize the software induced jitter at the application, operating system and device driver. For this purpose the transmitting application, besides writing the message in the transmission buffer, would also activate a timer with the message transmission instant. The management of the transmission buffer is of utmost importance to enforce the timely transmissions, as messages might have different priorities. It should also be possible to remove messages from the transmission buffer to cope, for example, with outdated messages.

Other requirements include a modular architecture to ease the modification of some building blocks or the addition of new ones and possibility to violate the inter-frame space so that the transmission of a sequence of time-triggered messages cannot be compromised by an event-triggered message. This latter aspect is of utmost importance for supporting deterministic MAC protocols, as it will then be possible to compute an upper limit for the time a message has to wait to be transmitted.

2. Design

The design of the ETSI ITS-G5 station was focused on the PHY and MAC layers as well as in the development of applications for demonstration purposes. Depending on the role, or target use (OBU or RSU), different requirements and constraints are applicable. However, both types of nodes require two 802.11p radio interfaces to be compliant with ETSI ITS-G5.

RSUs typically require a backhaul link, either wired or wireless, for connection to the Internet and remote access for management. A 3G/4G connection can be used if a wired connection is not available or not economically viable. Besides RSUs are installed in fixed locations, a GPS receiver is useful for time synchronization.

On the other hand, OBUs could provide a user interface for displaying information and warning messages to the driver. In this case a GPS receiver is useful for location and timing synchronization purposes. A 3G/4G network can be used as a backup connection for some OBU applications and it is also useful to support remote management and system upgrade operations. Additionally, they could have access to the OBD-II interface for getting vehicle status information.

2.1 Hardware-Software Partition

In terms of implementation, more precisely for hardware-software partition of the system, due to their complexity and operating system support, higher stack layers are more appropriate for software implementation, while lower layers require a hardware implementation to achieve the required performance and power efficiency. The main question is where to split the implementation, in order to obtain the best cost, performance and flexibility tradeoff.

The physical (PHY) layer performs several digital (de)modulation, synchronization, equalization and other signal processing operations that are more adequate for digital hardware implementation. Furthermore, digital-analog and baseband-RF conversions, as well as filtering and amplification stages require dedicated hardware, typically included in the RF frontend.

At the MAC layer, there are operations that are very time sensitive, such as, sensing the medium (e.g. based on clear channel assessment), medium access arbitration (e.g. based on the back-off mechanism), RSSI processing and start of message transmission. These operations require clock cycle accuracy and a close interaction with the PHY layer. Consequently they are more suited for hardware implementation. On the other hand, some more complex, but also less time critical MAC operations, such as, packet scheduling, message queue and QoS management are more appropriate for software implementation. This hardware-software partition is adopted widely in the implementation of communication systems and leads to a common division of the MAC layer: the Lower MAC (LMAC) implemented in hardware and the Upper MAC (UMAC) implemented in software. Moreover, the UMAC operations for common protocols are often implemented in some operating systems (e.g. Linux), which allows its direct use if the required LMAC primitives and interfaces are implemented in the specific hardware adapters or modems.

The layers above MAC are implemented in software to benefit from the common frameworks and operating system services. Moreover, during the initial market penetration phases, when only new vehicles come with embedded OBUs, DSRC connectivity can be provided to legacy vehicles in the form of add-on boxes and smartphones can play an important role in the graphical user interface.

2.1.1 ASIC vs. FPGA

Due to the close interaction and requirements, the LMAC and the digital PHY sub-layers are typically implemented on the same platform. For this purpose two different approaches or technologies can be adopted: ASIC or FPGA. The former provides better figures of merit in terms of operating frequency, power consumption and implementation resources but it is only economically viable for high production volumes. Furthermore, it is less flexible because once the design and manufacturing are finished, it is not possible to perform circuit modifications and a new design cycle is very time consuming and expensive. On the other hand, FPGAs have a much more flexible approach, with faster design cycles and lower non-recurring engineering (NRE) costs. They allow the implementation of the complete digital PHY processing chains using a SDR approach, as well as its integration with the custom LMAC logic. Both digital PHY and LMAC can be modified or extended easily by simply resynthesizing the design and reprogram the device.

2.1.2 FPGA Embedded Processor vs. Standalone SBC

On the other hand modern FPGA devices allow the integration of custom logic with standard processing blocks, in the form of hardwired or synthesizable IP cores, such as, microprocessors, memories, communication or protocol controllers, etc., leading to the concept of FPGA-based programmable SoC. Current FPGA devices and development frameworks provide good support to implement custom IP modules, integrate them with configurable microcontroller blocks and write the respective software, either in a standalone configuration or based on an embedded operating system (typically a derivative of a Linux distribution). This possibility currently allows the full integration from the application to the digital PHY in a single device, which seems to be appealing for this project. However, at early project stages, when most design decisions were taken, the embedded operating systems supported in FPGA-based programmable SoCs were more restricted and less supported versions of the main Linux distributions, which could impose some problems in the development of the ITS-G5 station layers. Thus, in the development of the ITS-G5 station, one preferred to use a separate Single Board Computer (SBC) for UMAC and upper layers implementation and keep the LMAC and digital PHY in a plain and lower cost FPGA device. This approach presents the following advantages:

- Distinct SBCs can be used for the OBU and RSU nodes with different processing capabilities and cost constraints. Moreover, multiples PHY devices and implementing multiple radio interfaces can be handled by a single SBC with the adequate processing capabilities.
- The number of SBCs commercially available is huge, supporting different operating systems (mainly Linux distributions) and processor architectures promotes the independence from the manufacturer. This is fundamental for lowering costs and simplifying product updates, since it is possible to individually upgrade the SBC, the FPGA module, or both. Moreover, most digital PHY and LMAC implementation is based on vendor independent VHDL code, which simplifies the migration between different FPGA families from the same or different vendor.
- The broader range of Linux distributions available for SBCs provided better support for device drivers and development tools than special distributions tailored for FPGA embedded processors. Additionally, the most recent Linux kernel advances are first provided for the mainstream processors used in SBC boards.
- Network support, including complete stack implementations is also an important advantage of most SBCs. In the particular case of the ITS-G5 station, the LMAC+PHY module implemented on FPGA acts as a conventional network adapter connected to the SBC through a standard USB port.
- Finally the GPS receiver, the 3G/4G modem and the OBD-II adapter can be attached directly to the SBC, since most of these devices provide drivers for major Linux distributions.

In brief, based on the above arguments, the designed ITS-G5 station is physically composed of the SBC (implementing the UMAC and upper layers), the FPGA module (implementing the LMAC and digital PHY layers), the RF frontend (including the AD/DA conversion and RF blocks) and a smartphone for graphical user interface. However, it is important to note that in future implementations it is worth to consider the use of an FPGA-based programmable SoC, such as Xilinx Zynq or Altera SoC, with a Hard Processor System (HPS) based on an embedded standard ARM Cortex core, which are nowadays fully supported by the Linux kernel and main distributions.

2.2 Architecture

Based on the previous arguments, a general architecture for the proposed IEEE WAVE/ETSI ITS-G5 station was defined. This new station, named

IT2S platform, is composed of the main blocks and interconnections presented in Fig. 3. There are three key physical components, namely the IT2S board, the Single Board Computer (SBC) and the Smartphone, which are responsible for executing specific tasks and implementing different layers of the protocol stack. The communication between the Smartphone, the SBC and the IT2S board is ensured through USB links. Furthermore, there are external interfaces to systems outside the IT2S platform, such as the On-board diagnostics (OBD- II) system available in all recent vehicles. The remaining external connections include the GPS and the DSRC 5.9 GHz antennas in the IT2S board, the Ethernet ports in the SBC and the 3G/4G interfaces in the Smartphone.

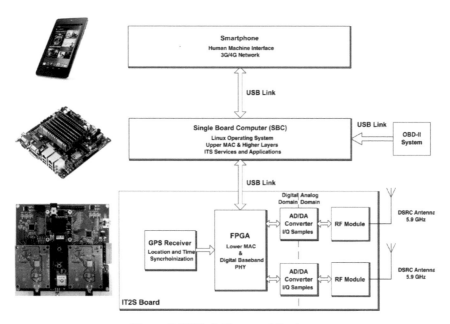

Figure 3: IT2S platform architecture.

The IT2S platform can operate either as an RSU or as an OBU. When the platform is working as an RSU, there is no need for a graphical user interface, and therefore the Smartphone is not included in the overall architecture of the system. Moreover, if there are several RSUs placed along the road, the network operator will want to remotely access those units or to make them working cooperatively. As a result, a back-hauling network will be required to exchange information among RSUs or between an RSU and a gateway node. In the case of IT2S platform, the Ethernet interface available in the Single Board Computer can be used for this purpose.

2.2.1 Smartphone

The Smartphone is responsible for implementing the graphical user interface with the driver and passengers of the vehicle. For instance, it should be able to display warnings in case of road accidents or traffic congestions. Another important advantage of the Smartphone is its capability to provide connectivity between the IT2S platform and a 3G or 4G network. This feature allows the remote diagnostics and access to the information available in the platform, as well as a possible upgrade of the software and the reconfiguration of the bitstream in the FPGA. Finally, it could also enable the implementation of the eCall service, which basically consists of an automatic call to the emergency number 112 in the event of a serious road accident.

2.2.2 Single Board Computer

As already mentioned, the main function of the Single Board Computer (SBC) is to execute the higher layers of the WAVE/ITS-G5 protocol stack, namely from the high level functionalities of the MAC layer, called Upper MAC (UMAC), to the Application layer. The SBC is a COTS embedded PC (e.g. Aaeon EMB-CV1) that runs a Linux-based operating system, providing a high degree of flexibility and more control over the operation of the system. When operating as an OBU, the SBC can also interact with the OBD-II system available in all recently manufactured vehicles. This way, it can access detailed information about the status and performance of the vehicle.

2.2.3 IT2S Board

The IT2S board is the core module of this platform, since it is not based on a commercial available solution, but in a device developed completely from scratch at Instituto de Telecomunicações (Aveiro site). It implements the recent IEEE 802.11p standard, focused on the MAC and physical layers of the protocol stack. However, since the implementation of the MAC layer resides in a hardware/software partition co-design, only the low level functionalities are executed in the IT2S board by the FPGA. This sub-layer that comprises the time-critical and deterministic operations of the MAC scheme, is designated as Lower MAC (LMAC).

The physical layer is implemented completely in the IT2S board, being divided in two main parts: the Analog and the Digital PHY. The Analog PHY is responsible for the signal processing operations in the analog domain, such as the up and down conversion from baseband to RF and vice-versa, respectively. On the other hand, the Digital PHY deals with signals in the digital domain, implementing the OFDM transmission and reception chains, converting bytes from a MAC frame into baseband In-phase and Quadrature (I/Q) samples and the reverse operation.

In order to cope with the simultaneous multi-channel operation requirement, the board includes two complete sets of hardware units (2 DSRC antennas and RF modules, 2 AD/DA processors and 2 digital PHY and LMAC modules inside the FPGA) for the implementation of the IEEE 802.11p standard in both radios. The IT2S board also incorporates a GPS receiver for location and synchronization purposes. The interconnection with the SBC is established through an USB link and it is based on a time multiplexing scheme, allowing the co-existence of several independent channels for accessing each radio unit separately, retrieving information from the GPS device, performing updates on the FPGA bitstream, etc.

2.3 Digital PHY

The design of the digital PHY encompasses the modulation and demodulation process of OFDM frames in baseband. According to the IEEE 802.11p standard, eight different modulation techniques can be used, as shown in Table 1. Nevertheless, only three out of these eight are considered mandatory, namely BPSK, QPSK and 16-QAM with coding rate 1/2. Hence, in order to be standard-compliant, an IEEE WAVE/ETSI ITS-G5 station has to meet all the radio specifications and regulatory requirements defined for these three modulation techniques.

Table 1: Modulation techniques and corresponding bit rates [5]

Modulation	Coding rate	Data rate (Mbps)
BPSK	1/2	3
BPSK	3/4	4.5
QPSK	1/2	6
QPSK	3/4	9
16-QAM	1/2	12
16-QAM	3/4	18
64-QAM	2/3	24
64-QAM	3/4	27

The digital PHY should also be able to support multi-channel access to the wireless medium, allowing the platform to operate in different channels over time. This implies a dynamic adjustment of the central frequency for the up/down-conversion processes, which is performed by the FPGA through the configuration of the corresponding hardware modules. As a result, the station can be tuned (not simultaneously, but at different time intervals) to all wireless channels allocated for the vehicular communication purposes (Fig. 2). This way, the platform can implement the several multi-channel schemes proposed in the IEEE 1609.4 standard [9] from the IEEE WAVE protocol stack.

2.4 Analog PHY

The design of the Analog PHY must also fulfill several requirements specified in the IEEE 802.11p standard, regarding the different power classes of operation, the transmit spectrum masks, the radio sensitivity etc.

For example, with respect to the four distinct power classes, the maximum transmit power levels at the antenna output are presented in Table 2. The table also shows the maximum values allowed for the Equivalent Isotropically Radiated Power (EIRP), which also takes into account the gain of the antenna, by summing up the output power of the transmitter with the maximum antenna gain.

Table 2: Maximum transmit power and EIRP for different power classes [5]

Power class	Maximum transmit power level (dBm)	Maximum allowed EIRP (dBm)
A	0	23
B	10	23
C	20	33
D	28.8	33 for non-government 44.8 for government

Another important requirement of the standard concerns the amount of power transmitted at frequencies beyond the necessary bandwidth. Hence, a transmit spectrum mask is defined with the maximum power levels for the spectrum occupied by the RF signal. A generic spectrum mask is presented in Fig. 4, while the power spectral density limits for different frequency offsets are specified in Table 3. These limits vary significantly

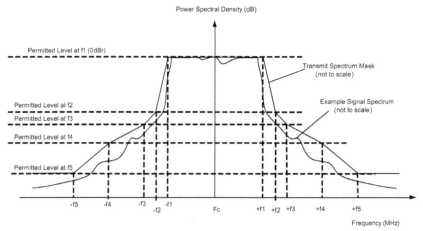

Figure 4: Generic transmit spectrum mask [5].

according to the power class the system uses to operate, putting higher demands in classes with higher transmit power levels.

Table 3: Maximum allowed power spectral density levels [5]

Power class	Maximum power spectral density (dBr)				
	±4.5 MHz (±f1)	±5.0 MHz (±f2)	±5.5 MHz (±f3)	±10 MHz (±f4)	±15 MHz (±f5)
A	0	-10	-20	-28	-40
B	0	-16	-20	-28	-40
C	0	-26	-32	-40	-50
D	0	-35	-45	-55	-65

On the receiver side, the radio sensitivity defines the minimum amount of power required to achieve a predefined value of packet error rate. Table 4 exhibits the minimum sensitivity levels required by the standard for all possible modulation techniques.

Table 4: Minimum sensitivity values for all modulations [5]

Modulation	Coding rate	Minimum sensitivity
BPSK	1/2	-85
BPSK	3/4	-84
QPSK	1/2	-82
QPSK	3/4	-80
16-QAM	1/2	-77
16-QAM	3/4	-73
64-QAM	2/3	-69
64-QAM	3/4	-68

3. Implementation

This section presents the implementation aspects of the different parts of the architecture proposed for the IEEE 802.11p station [10]. Hence, a detailed description of the analog and digital PHY, as well as of the modules that compose the MAC layer, is provided. As already explained, most of the physical and MAC layers were implemented in the IT2S board, including the LMAC and both the digital and the analog PHY. Only the UMAC sub-layer was developed in the Single Board Computer. Figure 5 presents a picture of the IT2S board with the identification of its main building blocks.

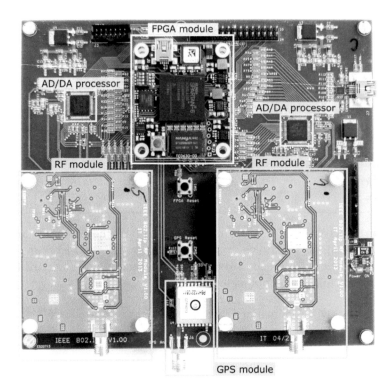

Figure 5: IT2S board.

3.1 Analog PHY

The analog part of the physical layer is essentially constituted by the RF module of the IT2S board. Figure 6 presents the RF module architecture

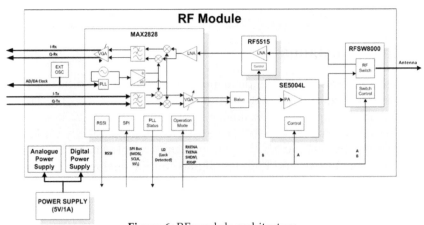

Figure 6: RF module architecture.

with its essential components. The main function of this module is the conversion and amplification of the I/Q analog signals from baseband to an RF signal in the 5.9 GHz frequency band and vice-versa.

3.1.1 Main Building Blocks

The RF module is mainly constituted of four different components: an RF switch, a low noise amplifier, a power amplifier and an RF transceiver integrated circuit.

- The RF switch is responsible for commuting the high frequency signal between the reception and the transmission paths, depending in which one of these states the IT2S board is operating at a determined instant.
- The low noise amplifier enables the amplification of the received RF signal when the RF module is in the reception mode.
- The power amplifier is responsible for the amplification of the RF frequency signal before it reaches the air interface, when the module is in the transmission mode.
- The transceiver circuit implements the frequency up-conversion from baseband to RF when the system is transmitting a message, and the down-conversion from RF to baseband when the platform is in the receiving state.

There is also some additional circuitry utilized for signal conditioning purposes, in order to adapt the baseband signal levels that interconnect the AD/DA processor and the transceiver module.

3.2 Digital PHY

The digital physical layer of the IT2S board was implemented completely in the FPGA, allowing fast prototyping as well as flexible and reconfigurable development of the required hardware blocks and control modules. Figure 7 depicts a simplified overview of the digital PHY architecture. Basically it consists of five main modules: the MAC Interface + PHY Controller, the Hardware Controller + CCA Mechanism, the transmission chain, the reception chain and the AD/DA Interface module.

The first one is responsible for the interconnection between the MAC and the PHY layers, according to what is defined in IEEE 802.11 standard. It deals with the packet delivery in both directions between these two layers and controls the remaining blocks of the physical layer by switching them between transmission and reception modes.

The main function of the hardware controller block is to configure the AD/DA processor and the RF module dynamically, in accordance with the operating parameters specified by the upper layers of the protocol stack, e.g. the frequency channel. This module is also responsible

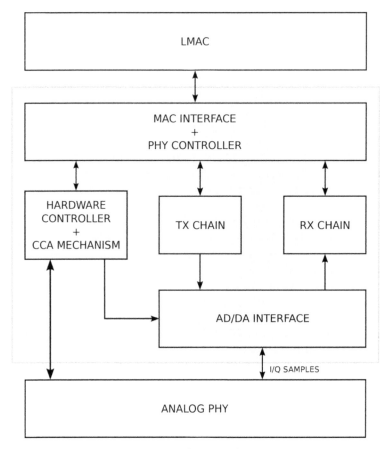

Figure 7: Digital PHY architecture.

for implementing the Clear Channel Assessment procedure and the Automatic Gain Control mechanism.

The transmission chain block performs a sequence of signal processing operations, as presented in Fig. 8, in order to implement a physical transmitter based on the OFDM modulation scheme as defined by the standard. It takes as an input the transmission parameters (e.g. the bitrate) and the frame from the MAC layer and converts it to I/Q baseband samples that will be delivered to the Digital-to-Analog Converters (DACs).

The reception chain implements the inverse operations of the transmission chain, by taking the samples from the Analog-to-Digital Converters (ADCs), demodulating them into valid OFDM symbols and producing the bytes that constitute a MAC frame (Fig. 9). If an error is detected during the demodulation process inside the reception chain, no frame is delivered by this unit, but the event is reported to the MAC-PHY controller module that will forward it to the MAC layer.

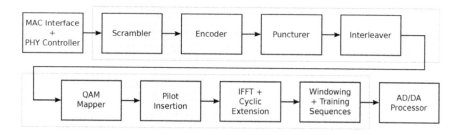

Figure 8: Block diagram of the digital transmission chain.

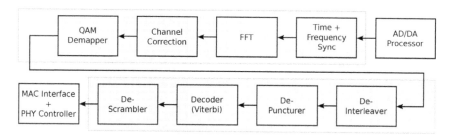

Figure 9: Block diagram of the digital reception chain.

Finally, the AD/DA Interface module guarantees correct data exchange between the AD/DA processor and the FPGA, by taking care of the bus access from both transmission and reception chains.

3.3 MAC Layer

As previously discussed in section 2, it was decided, during the design phase, to implement the MAC layer in a hardware/software partition approach [11]. In brief, this results in the decomposition of the MAC layer into the LMAC and UMAC. The LMAC is implemented in hardware (FPGA) and interacts directly with the physical layer. It is responsible for the execution of the time critical functionalities of the MAC layer that require a high temporal accuracy and predictability. On the other hand, the UMAC deals with the less time constrained MAC operations, being responsible for the interface with the higher layers of the protocol stack. This sub-layer is completely defined in software and it is implemented in the embedded computer. The interface between the UMAC and the LMAC is established through an USB connection between the FPGA and the embedded PC. The global architecture of the MAC layer and the interconnection with the remaining layers of the protocol stack can be observed in Fig. 10.

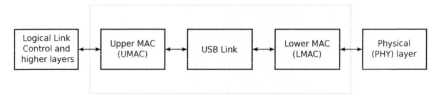

Figure 10: Global overview of the MAC architecture and interfaces.

3.3.1 Lower MAC

The main functions of the Lower MAC (LMAC) sub-layer are:

- manage PHY interface: channel change and frame transmission and reception
- provide memory buffers to store frames
- generate and verify the Frame Check Sequence (FCS) for each frame
- implement the Carrier Sense Multiple Access with Collision Avoidance (CSMA/CA) mechanism
- provide suitable interface for the Upper MAC (UMAC)

The LMAC architecture is presented in Fig. 11. There are two modules interacting directly with the USB communication system: the Command Processor and the Event Conveyor. The first is responsible for decoding and processing the UMAC commands, while the second conveys the LMAC

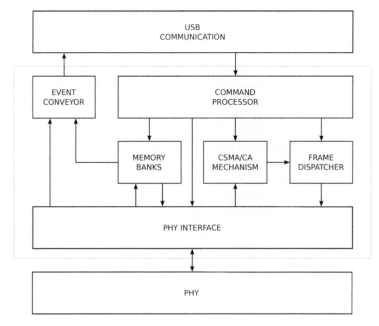

Figure 11: LMAC architecture and interfaces.

events and the received frames to the UMAC. Both modules are connected with the Memory Banks, where all the transmitting and receiving frames are stored. The CSMA/CA Mechanism implements the method defined in the standard to provide multiple access to the wireless medium. The Dispatcher is responsible for implementing the various priority queues and issue the packet transmission requests whenever possible, according to the CSMA rules. Finally, the PHY Interface module makes available to the LMAC an abstraction layer of the PHY, by providing an adequate set of signals to be used by all other modules.

3.3.2 Upper MAC

The Upper MAC (UMAC) sub-layer is responsible for:

- interface with the higher layers of the protocol stack, namely the Logical Link Control (LLC)
- build the header of the MAC frames for transmission
- analyze the header of the received frames
- provide suitable interface for the Lower MAC (LMAC)

The UMAC architecture is composed by several separate processes, as presented in Fig. 12. The role of the Request Scheduler is to prioritize the transmission of the different command requests to the LMAC through the USB communication system. There are three different types of commands that could be issued to the LMAC. These could be related with

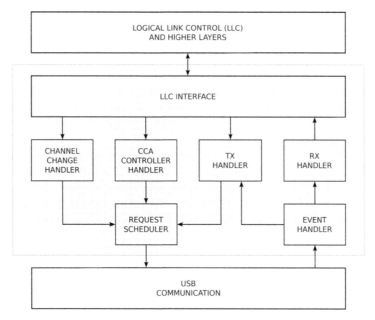

Figure 12: UMAC architecture and interfaces.

the transmission of a frame (Tx Handler), the configuration of the Clear Channel Assessment (CCA) parameters (CCA Controller Handler) or the channel change (Channel Change Handler). The Event Handler process is responsible for analyzing the type of event conveyed by the LMAC and forward it to the appropriate handler. It could be a transmission event that will be forwarded to the Tx Handler, in order to allow the maintenance of accurate information regarding the state of the LMAC memory banks, or it could be related with the reception of a new frame, which will be managed by the Rx Handler. At last, there is a LLC Interface process that is responsible for communicating with the higher layers of the protocol stack.

3.4 Verification Support

From the description above, it is obvious that an important part of the ITS-G5 system is implemented in digital hardware inside the FPGA. The test and verification of such complex FPGA based system is a challenging task, given the lack of flexibility that characterize the tools available for that purpose. While simulation tools provided by the vendor can be easily employed to verify the robustness of the system under controlled stimulus, there is a need for the development of hardware tools that allow that analysis under real operation scenarios.

A Spartan-6 FPGA from Xilinx was used in this project. As a result, the specific tool provided by the vendor for hardware testing and validation is the ChipScope Pro utility. Despite its usefulness in particular applications, this tool has some shortcomings when a complete and intensive system test is required. For example, the tool is only able to capture data from the FPGA, so there is no possibility to inject previously known signals into the system. Moreover, expensive and dedicated hardware is required to interact with the device, since the ChipScope tool needs a JTAG interface to access the FPGA. There are other disadvantages such as the need for resynthesizing the digital hardware design when the trigger signal or the capture point is modified.

For the given reasons, a custom FPGA test and verification tool named Wiretap was developed in the context of this project. This tool was designed with the initial goal of validating the operation of the digital transmission and reception chains, but then it was extended in order to allow the test of any generic hardware module inside the FPGA.

3.4.1 Wiretap

Wiretap is a tool to perform systematic capture and injection of data samples at any point of the FPGA design. It also allows the online configuration of specific system parameters, necessary to apply some modifications in the design of the corresponding hardware modules.

Wiretap is fully integrated with Matlab, only requiring, for communication with the PC, a USB or UART module in the development board. In the IT2S board both options are available: either the UART interface can be used or a dedicated USB endpoint could be assigned to allow simultaneous operation with the remaining USB communications.

A global overview of the Wiretap integration with the IT2S board is presented in Fig. 13. Several injection/capture points could be inserted in the design of the system and then one of them could be dynamically selected in the MatLab through a multiplexing/demultiplexing mechanism. In addition to this, any hardware module could be configured through the Wiretap instantiation in the FPGA, by receiving commands with the specified parameters from MatLab.

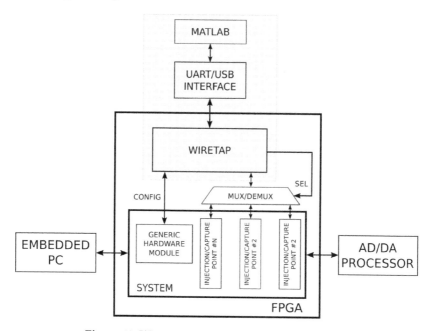

Figure 13: Wiretap integration with the IT2S board.

Since Wiretap is integrated with MatLab, the captured data can be immediately analyzed and the injection signals can be easily generated. These characteristics enable fast validation of the design and quick development of the project under test. In the following list, some use case scenarios of the Wiretap tool during the verification phase of the system are presented:

- inject a well formatted OFDM baseband sequence at the input of the reception chain, in order to validate the demodulation process of a frame;

- capture the samples being read by the ADC, in order to test, for instance, adjacent channel interference;
- capture the sequence of samples corresponding to a complete IEEE 802.11p frame, at the input of the reception chain or at the output of the transmission chain, thus allowing the comparison between the results produced by the hardware implementation and by the functional MatLab model.

4. Validation

This section describes the validation procedure of the developed IT2S platform. It presents the test environment with the experimental setups used to evaluate the most important metrics of the platform in the transmission and reception modes, as well as the obtained results. At the end, the FPGA resource usage is presented, showing that there is enough space available for the implementation of possible extensions to the IEEE WAVE/ETSI ITS-G5 standard.

4.1 Test Environment

The IT2S platform was tested in a laboratory environment, in order to evaluate its compliance with the requirements defined in the IEEE 802.11 standard. Several metrics were analyzed, leading to the preparation of two different experimental setups, one for the evaluation of the transmission metrics and the other for the receiver tests.

The experimental setup used to evaluate the performance of the IT2S platform in transmission mode is depicted in Fig. 14. The embedded PC control the transmission of frames to the IT2S board, which forward them to a Vector Signal Analyser (VSA), after proper modulation and upconversion to RF (in the 5.9 GHz frequency band). Then, at the end of each test, the results are automatically collected from the VSA (R&S®FSQ). A picture of this experimental setup under test in the laboratory environment is shown in Fig. 15.

For the receiver tests, the experimental setup depicted in Fig. 16 was employed. A Vector Signal Generator (R&S®SMU200A) is configured to transmit frames to the IT2S board, which after the downconversion and demodulation processes, sends them to the embedded PC. Figure 17 presents the actual setup being tested in the laboratory.

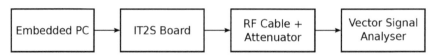

Figure 14: Block diagram of the setup used in the transmitter tests.

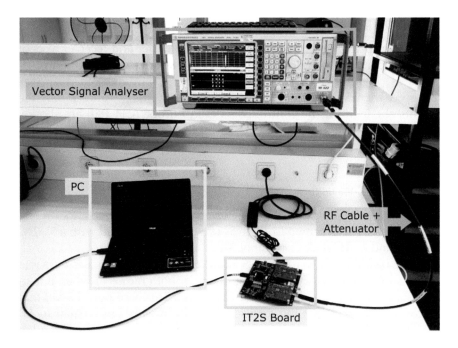

Figure 15: Experimental setup for transmitter tests in the laboratory.

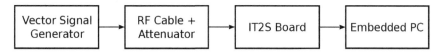

Figure 16: Block diagram of the experimental receiver tests.

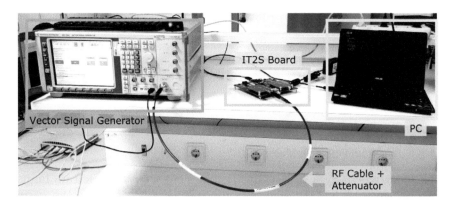

Figure 17: Experimental setup for receiver tests in the laboratory.

4.2 Experimental Results

The experimental results obtained in the transmitter and receiver tests of the IT2S platform are presented next. These tests were performed in all wireless channels reserved for the vehicular communication purposes, but since there was practically no variation with this parameter, only the results obtained for the control channel in Europe (CH 180) are shown. Similarly, only the results corresponding to the modulations considered mandatory by the standard are presented (BPSK, QPSK and 16-QAM with coding rate 1/2).

4.2.1 Energy Consumption

The energy consumption of the IT2S board was analyzed in several operating states, as shown in Table 5. The board is supplied with a 5V DC voltage and a maximum current of 3A.

The results show that the IT2S board consumes more or less the same average power when it is receiving in one or both radios or when it is in idle state (neither transmitting nor receiving). This base power level that is consumed whenever the device is turned on, is roughly equal to 4.3 W. The consumed power increases significantly when the IT2S board is transmitting at maximum power. As can be observed in Table 5, a radio operating in transmission mode adds approximately 2 W to the total amount of power consumed by the system (6.35 W and 8.55 W when one or both radios are transmitting at maximum power, respectively).

Table 5: Energy consumption of the IT2S board

Operating states	Voltage (V)	Current (mA)	Average power (W)
Idle state		856	4.28
Receiving (1 radio)		878	4.39
Receiving (2 radios)	5	890	4.45
Transmitting (1 radio)		1270	6.35
Transmitting (2 radios)		1710	8.55

4.2.2 Spectrum Emission Masks

To verify whether the IT2S board is compliant with the spectrum limits defined in IEEE 802.11p standard, the spectrum occupied by the radio in transmission mode has to be compared with the applicable transmit spectrum mask. Figure 18 shows the power spectral density of the signal

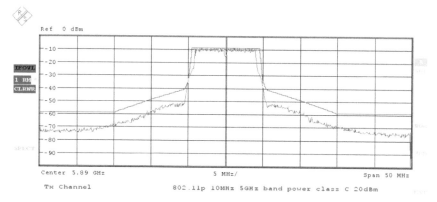

Figure 18: Spectral mask and signal emitted by the platform (power class C).

emitted by the IT2S board when transmitting with an output power of 20 dBm (maximum value for operation in power class C). It can be observed that the signal transmitted by the platform (the blue line in the figure) is inside the spectrum mask defined by the standard for the power class C (represented by the red line).

After testing the operation of the IT2S board in transmission mode with different power levels, the obtained results attest that the platform is compliant with power classes A (0 dBm), B (10 dBm) and C (20 dBm). However, it could not fulfil the high demands posed by the operation in power class D (28 dBm), due to the non-linearities of the power amplifier when transmitting with high power levels.

4.2.3 Demodulation-Constellations and EVM

Another important parameter used to evaluate the transmitter performance is the Error Vector Magnitude (EVM) parameter, measured in the I/Q constellation symbols. An example of the constellation observed in the Vector Signal Analyzer for the BPSK modulation is shown in Fig. 19. The results associated with this measurement, namely the EVM metric, can also be visualized in the VSA (Fig. 20).

In order to evaluate the EVM parameter, a batch of 10,000 packets with random content and a size of 1000 bytes, was sent in each test. Additionally, three different power levels (0, 10 and 20 dBm), corresponding to the maximum values for the power classes A, B and C, were analyzed.

The measured EVM values are presented in Table 6. From these results, it can be concluded that the IT2S board is inside the EVM limits defined by the IEEE 802.11p standard.

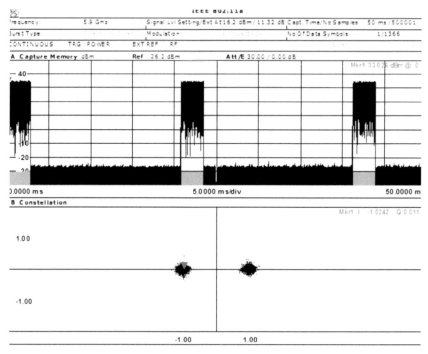

Figure 19: I/Q constellation for the BPSK modulated packets.

IEEE 802.11a					
Frequency 5.9 GHz	Signal Lvl Setting/Ext At16.2 dBm / 11.32 dB Capt. Time/No Samples 50 ms / 500001				
Burst Type	Modulation		No Of Data Symbols 1/1366		
CONTINUOUS TRG POWER	EXT REF RF				

Result Summary					
	Min	Mean	Limit	Max	Limit	Unit
EVM All Carriers	11.24	11.40	56.23	11.67	56.23	%
	- 18.98	- 18.86	- 5.00	- 18.66	- 5.00	dB
EVM Data Carriers	11.24	11.41	56.23	11.71	56.23	%
	- 18.98	- 18.86	- 5.00	- 18.63	- 5.00	dB
EVM Pilot Carriers	11.11	11.39	39.81	11.85	39.81	%
	- 19.08	- 18.87	- 8.00	- 18.53	- 8.00	dB
IQ Offset	- 16.51	- 16.50	- 15.00	- 16.49	- 15.00	dB
Gain Imbalance	- 0.13	- 0.03		0.05		%
	- 0.01	- 0.00		0.00		dB
Quadrature Error	0.10	0.12		0.14		°
Center Frequency Error	49782.14	49784.99	∗ 118000	49787.56	∗ 118000	Hz
Symbol Clock Error	8.43	8.45	∗ 20	8.46	∗ 20	ppm
Burst Power	27.13	27.14		27.14		dBm
Crest Factor	7.95	8.06		8.18		dB

Figure 20: EVM measurements for the BPSK modulated packets.

Table 6: EVM measurements for three different power classes

Modulation	Coding rate	Transmit power level (dBm)	Measured EVM$_{rms}$	Maximum EVM$_{rms}$ (%)
BPSK	1/2	20	13.26	
		10	14.72	56.23
		0		
QPSK	1/2	20	11.79	
		10	13.63	31.62
		0	13.87	
16-QAM	1/2	20	11.55	
		10	12.14	15.85
		0	12.32	

4.2.4 Sensitivity

In the receiver tests, the ability of the IT2S board to correctly receive and demodulate an incoming signal was verified. These tests were focused specially on the Packet Error Rate (PER) measurements at very low and very high input power levels. The PER is defined as the ratio between the number of incorrectly received packets, either undetected or received with errors, and the total number of transmitted packets. According to the standard, the receiver should guarantee a PER below 10% for high input power levels (signals with at least -30 dBm). The obtained results proved that the platform complies with this requirement for all mandatory modulations.

Similarly, the IEEE 802.11p standard defines a maximum value of 10% for the PER at very low power values. The sensitivity of the system corresponds to the minimum input power level, at which the PER is less or equal than 10%, for packets with 1000 random bytes. The results of the sensitivity measurements are displayed in Table 7. These results are not compliant with the minimum sensitivity values defined in the standard. For example, there is a gap of 10 dB between the measured sensitivity and the required value for the BPSK modulation. This limitation is related to the amount of noise introduced by the analog components of the reception path.

Table 7: Sensitivity measurements of the IT2S board

Modulation	Coding rate	Measured sensitivity (dBm)	Minimum sensitivity (dBm)
BPSK	1/2	-75	-85
QPSK	1/2	-73	-82
16-QAM	1/2	-72	-77

As a consequence, a new version of the IT2S board has been developed in order to solve this issue and the non-conformance with the operation in power class D. Preliminary results have shown that this improved design meets the sensitivity requirements imposed by the standard, as can be observed in Table 8.

Table 8: Sensitivity measurements of the improved IT2S board

Modulation	Coding rate	Measured sensitivity (dBm)	Minimum sensitivity (dBm)
BPSK	1/2	-91	-85
QPSK	1/2	-88	-82
16-QAM	1/2	-82	-77

4.3 FPGA Resource Usage

To conclude this section, Table 9 presents the total number and the percentage of resources used in the FPGA Spartan-6. These results take into account the implementation of two Lower MAC + digital PHY sub-layers, one for each radio, together with the integrated Wiretap tool. It can be observed that the hardware logic blocks are not totally occupied and there are still resources available for the implementation of possible extensions to the WAVE/ITS-G5 standards, as described in greater detail in the next section.

Table 9: IT2S platform – Resource usage on Spartan-6 LX150 FPGA

Logic resources	Used	Total	Percentage used
Flip Flops	39292	184304	21%
LUTs	38111	92152	41%
RAMB16BWERs	196	268	73%
RAMB8BWERs	48	536	8%
DSP48A1s	80	180	44%

Logic max. frequency – 158.188 MHz

5. ITS-G5 Extensions

Having a white box access to a ITS-G5 platform opens an array of possible extensions and improvements beyond the standard, while maintaining compatibility at the physical and MAC layers, whenever required. Examples of such improvements are support for deterministic MAC protocols and fault tolerance mechanisms.

5.1 Support for Deterministic MAC Protocols

Real-time dependable communication requires deterministic transmission instants and bounded latency for inter-vehicular communication. Carrier Sense Multiple Access (CSMA) has serious drawbacks in what concerns to the timeliness of the communications, since it allows collisions to take place in the channel and it may cause unbounded delays before channel access. Support for deterministic MAC protocols requires a set of functionalities, which are not usually available in COTS transceivers, both at hardware and software level.

Detailed channel load measurement is required, in order to allow for decentralized solutions where the stations adequate their transmission patterns to the medium occupation. It is also necessary to capture detailed channel usage statistics, which should include fine grained RSSI measurements. Another required functionality is to be able to execute run time modifications on the sensitivity level, above which the medium is regarded as busy and therefore packets are detected. This functionality would allow the implementation of mechanisms such as Decentralized Congestion Control (DCC), where under heavy load conditions the minimum signal level for busy medium declaration is raised. This limits the device's effective range and reduces the coverage area and thereby the maximum amount of vehicles competing for the shared medium. This should be done during the inter-frame space, for not interfering with an ongoing message reception.

Disabling any collision avoidance mechanisms is required to be able to transmit packets in a timely fashion and with disregard for any ongoing transmission, resembling the single shot transmission mode present in some fieldbus protocols. Although this could create forced collisions, some proposed protocols explicitly require the behaviour in order to be able to perform strict Time-Division Multiple Access (TDMA). Having a TDMA based protocol is not enough to guarantee deterministic communications, as a node outside the protocol can corrupt the TDMA schedule by transmitting a message in a slot assigned to another node. A possible solution for this problem is giving all nodes participating in the deterministic protocol, the possibility of violate the Inter Frame Space (IFS). In this way, a node outside the deterministic TDMA-based protocol would sense the medium busy during the TDMA slots, and would be allowed to transmit only outside the reserved time windows.

The IT2S platform also needs to be tightly coupled to a time synchronization mechanism to support low jitter TDMA implementations. This support should be present in hardware to shield the time sensitive parts of the system from interference from the higher layers of the communication stack. That is, the transmission instant of a message should be independent from the time when it is written in the transmission buffer, to minimize the software induced jitter (application + operating system + device driver). For this purpose the transmitting application, besides writing the message in the transmission buffer, would also activate a timer with the message transmission instant. The management of the transmission buffer is of utmost importance to enforce the timely transmissions, as messages may have different priorities. It should also be possible to remove messages from the transmission buffer to cope, for example, with outdated messages.

5.1.1 Time-triggered Packet Transmission

Mechanisms to support time-triggered packet transmission were already implemented and validated [12]. For transmitting a packet the application stores it in the transmission buffer, together with the time instant when it should be dispatched. At the transmission instant, the dispatcher sends the packet to the transmission chain, disabling the backoff and EDCA procedures. This is the so-called single-shot transmission mode, where a packet is transmitted, with no re-transmissions, regardless of the state of the medium.

Experimental results [12], using two IT2S platforms and GPS for clock synchronization, indicated that the delay for transmitting a packet was 16.5 µs, with a jitter of 3.5 µs. This delay is independent from the packet size as one measured the time from the transmission time-stamp to the reception of the first bit in the reception chain of the receiving platform. Notice that this delay includes the delays of the transmission and reception chains plus the propagation time in laboratory conditions, with transmitter and receiver placed at a distance of 5 meters. These results are promising and in line with the requirements of closed-loop distributed control applications.

5.1.2 Violating the Inter-frame Space

Inter Frame Space is the time that a node needs to wait for before sensing the medium free and then to send the packet. To avoid collisions, an IEEE 802.11p compliant node running CSMA/CA based MAC protocols needs to wait IFS duration before initiating a transmission. The minimum defined IFS in IEEE 802.11p standard is 58 µs to provide enough time to the mobile nodes to communicate. Experimental results [13] indicate that the IFS can be reduced in the IT2S platform down to 17 µs.

5.2 Fault Tolerance Mechanisms

Wireless vehicular communications are targeted to safety applications, where failures may cause human and economic losses. Thus, fault-tolerance mechanisms need to be put in place to prevent faults to became failures. Several types of faults must be considered in vehicular scenarios.

For example, the wireless channel is regularly affected by transient faults in the communication link due to the rapidly changing radio propagation patterns and road traffic conditions. Furthermore, channel permanent faults could also occur, due to unregulated interference, which can be considered as malicious faults, since the spectrum band assigned for wireless vehicular communications is reserved by law. Not only the wireless channel, which is a single point of failure in vehicular systems, but also the nodes of the network should be regarded as a possible source of problems. For instance, hardware and software faults can affect the operation of either the road-side units (RSUs) that constitute the network infrastructure or the on-board units (OBUs) placed inside each vehicle. For these reasons, a careful design work must be performed in the deployment of vehicular communications systems, by taking into consideration the general dependability aspects of safety-critical applications, as well as the specific issues that arise in the vehicular context.

5.2.1 Fail-silence Enforcement

An important type of failure that must be considered in the design of fault tolerant vehicular communication systems is the babbling idiot failure mode. A babbling idiot node sends unsolicited messages at arbitrary points in time without respecting the media access rules. As a consequence, this faulty node can disable nodes with legitimate messages to access the network. This scenario is completely unacceptable and therefore a node should only exhibit simple failure modes and ideally it should have just a single failure mode, the fail silent failure mode, i.e., it produces correct results or no results at all. In this matter, a node can be fail-silent in the time domain, i.e., transmissions occur at the right instants, only, or in the value domain, i.e., messages contain correct values, only. With fail silence behaviour, an error inside a node cannot affect other nodes and thus each node becomes a different fault confinement region.

A fail silence enforcement mechanism has already been proposed [14], in order to avoid the transmission of messages with erroneous contents or at invalid instants. The operation of the proposed fail silence mechanism consists of an internal redundancy scheme, based on the replication of the IT2S platform. Two complete sets of single board computers and IT2S boards are used to produce messages, whose purpose is to be disseminated through the air medium. The fail silence enforcement entity, which is implemented in a separate FPGA, then compares the values and the timing of the messages produced by these two sets and, if everything is working correctly, it validates the frame and allows its transmission by one of the platforms. On the other hand, if the frames differ or are significantly out of phase, the entity silences the system until a restart signal is received.

5.2.2 Low-latency Station Replication

One of the most common methods to build fault-tolerant distributed systems is to replicate subsystems that fail in an independent way. The goal of this approach is to give other subsystems the idea that the delivered service is provided by a single entity. In order to enforce consistency among the replicas, there are two main categories of replication schemes: active and passive replication.

When compared to passive replication, active replication schemes have lower latency, since there is no need for an explicit recovery procedure when one of the replicas fails and the other ones will continue to provide correct service. The task of designing mechanisms that enforce node replication is facilitated if all nodes are fail-silent, since the only possibility for them to fail is by not issuing any message to the network.

To mitigate any communication discontinuity caused by a fail-silent faulty replica, the backup one should take over the operation as if the primary one has not failed, for example, not allowing a third node to transmit a message before the backup replica, upon failure of the primary, as this may compromise timeliness.

A solution to this problem is making the active replica transmit with an IFS of, e.g., 20 μs and the backup one with an IFS of 40 μs. Notice that both IFS are less than the minimum defined in the standard, so no other node would be able to transmit before these. This calls for a new transmission mode, in which the backup node needs to violate the IFS and sense the medium to assess if the primary replica has failed to transmit.

This way, if the active replica is able to transmit the message in the planned instant, the backup will sense the wireless medium as occupied and will conclude that the primary system is free of error. Hence, it will not issue any message to the air, avoiding any overlap with the active node. On the other hand, if at that moment the medium is perceived as free, the backup replica will continue the transmission of its message and replace the operation of the previously active one. As a result, the messages will still be transmitted on time, since only a small delay (lower than the standard IFS value) is introduced.

6. Conclusions

This chapter presents the development of a wireless communications platform for vehicular applications. It focuses specially on the description of the main design challenges and on the definition of a possible architecture for the implementation of the lower layers of the protocol stack, namely the physical and medium access control layers. The proposed station (IT2S platform) is based on the IEEE 802.11p standard, which constitutes the common basis for the development of the protocol stacks for vehicular communications in USA and in Europe, the IEEE WAVE and the ETSI ITS-G5 family of standards, respectively. The IT2S platform provides simultaneous multi-channel operation by implementing two complete sets of hardware and software blocks, which allows the station to have one radio dedicated to safety purposes and the other one tuned in an infotainment channel.

This platform can operate either as an RSU or as an OBU and is constituted by three key components: a Smartphone, a Single Board Computer and the IT2S board. The latter one implements the entire physical layer and the low level functionalities of the MAC layer (Lower MAC). The remaining functions of the MAC layer (Upper MAC) are executed in the Single Board Computer. This hardware/software MAC partition derives from the design choice to separate the timely and predictable operations of the MAC layer, performed by an FPGA in the IT2S board, from the high level ones, which are software implemented in the SBC. The physical layer implementation follows a software defined radio approach, in which part of OFDM modulation/demodulation processes are executed in the digital domain by an FPGA, while the components closer to the antenna operate in the analog domain. The obtained results show that the IT2S platform complies with the majority of the specifications defined in the IEEE 802.11p standard. Nevertheless, there are some requirements that cannot be met completely, namely the sensitivity levels and as a consequence a new version of the platform is being developed in order to overcome these issues.

Additionally, some extensions and improvements to the standard operation of vehicular communications nodes were presented, with the main goal of providing a more deterministic and dependable behaviour in safety-critical scenarios. For instance, supporting mechanisms for deterministic MAC protocols have already been implemented, together with fault tolerance techniques that improve the reliability of the system. These extensions are possible only by having a white box access to a ITS-G5 station, such as the case of IT2S platform.

Acknowledgements

This work is funded by National Funds through FCT – Fundação para a Ciência e a Tecnologia under the PhD scholarship ref. SFRH/BD/52591/2014 and the project PEst-OE/EEI/LA0008/2013, by the European Union's Seventh Framework Programme (FP7) under grant agreement n. 3176711 and by BRISA, under research contract with Instituto de Telecomunicações - Aveiro.

References

1. Commission, E. 2015. Statistics – accidents data. http://ec.europa.eu/transport/road safety/specialist/statistics/index en.htm, online, accessed on April 27 2015.
2. Bilstrup, K., E. Uhlemann, E.G. Ström and U. Bilstrup. 2009. On the Ability of the 802.11p MAC Method and STDMA to Support Real-Time Vehicle-to-Vehicle Communication. *EURASIP J. Wireless Comm. and Networking*, vol. 2009.
3. Eckhoff, D., N. Sofra and R. German. 2013. A performance study of cooperative awareness in ETSI ITS G5 and IEEE WAVE. In: *Wireless On-demand Network Systems and Services* (WONS), 10th Annual Conference on March 2013, pp. 196–200.
4. ETSI. 2011. Final draft ETSI ES 202 663 V1.1.0 (2009-11), ETSI Standard, Intelligent Transport Systems (ITS); European profile standard for the physical and medium access control layer of Intelligent Transport Systems operating in the 5 GHz frequency band.
5. IEEE Standard for Information Technology. 2012. Telecommunications and information exchange between systems. Local and metropolitan area networks–Specific requirements. Part II: Wireless LAN Medium Access Control (MAC) and Physical Layer (PHY) Specifications. IEEE Std 802.11-2012 (Revision of IEEE Std 802.11-2007), pp. 1–2793.
6. Subramanian, S., M. Werner, S. Liu, J. Jose, R. Lupoaie and X. Wu. 2012. Congestion Control for Vehicular Safety: Synchronous and Asynchronous MAC Algorithms. In: *Proceedings of the Ninth ACM International Workshop on Vehicular Inter-networking, Systems, and Applications*. ser. VANET '12. New York, NY, USA: ACM, pp. 63–72.
7. Autolitano, A., C. Campolo, A. Molinaro, R. Scopigno and A. Vesco. 2013. An insight into Decentralized Congestion Control techniques for VANETs from ETSI TS 102 687 V1.1.1. In: *Wireless Days (WD)*, 2013 IFIP, pp. 1–6.
8. Kenney, J. 2011. Dedicated Short-Range Communications (DSRC) Standards in the United States. *Proceedings of the IEEE*, vol. 99, no. 7, pp. 1162–1182.
9. IEEE Standard for Wireless Access in Vehicular Environments (WAVE). 2011. Multi-channel Operation. IEEE Std 1609.4-2010 (Revision of IEEE Std 1609.4-2006), pp. 1–89.
10. Almeida, J. 2013. Plataforma multi-rádio para comunicações veiculares DSRC 5.9 GHz. Master's thesis. University of Aveiro, Aveiro, Portugal.
11. Cruz, C. 2014. Support Mechanisms for Deterministic 802.11p MAC. Master's thesis. University of Aveiro, Aveiro, Portugal.
12. Cruz, C., J. Ferreira and A. Oliveira. 2015. Supporting Deterministic Medium Access Control in Wireless Vehicular Communications. In: *Proceedings of the IEEE 82nd Vehicular Technology Conference*. ser. VTC2015-Fall. IEEE.
13. Khan, A., J. Almeida, B. Fernandes, M. Alam, P. Pedreiras and J. Ferreira. 2015. Towards Reliable Real Time Vehicular Communications. In: *International Workshop on Cooperative Sensing for Smart Mobility* (COSSMO) of the 18th International IEEE Conference on *Intelligent Transportation Systems* (ITSC 2015).

14. Ferreira, J., A. Oliveira, J. Almeida and C. Cruz. 2013. Fail silent road side unit for vehicular communications. In: *SAFECOMP 2013 – Workshop ASCoMS* (Architecting Safety in Collaborative Mobile Systems) of the 32nd International Conference on *Computer Safety, Reliability and Security*, Toulouse, France.

When Buses Become Smart: The OBIT Experience

Matteo Petracca[1,*], Luca Maggiani[1], Stefano Bocchino[1]
Claudio Salvadori[2] and Luciano Niccolai[2]

[1]Institute of Communication, Information and Perception Technologies
Scuola Superiore Sant'Anna, Via Giuseppe Moruzzi 1, 56124, Pisa, Italy
[1]Scuola Superiore Sant'Anna Research Unit National Inter-University
Consortium for Telecommunications, Via Giuseppe Moruzzi 1, 56124, Pisa, Italy
[2]New Generation Sensors Srl, Via Gioacchino Volpe 12, 56121, Pisa, Italy

Abstract

In the last several years Intelligent Transport Systems (ITSs) have become
more and more effective in providing advanced services to all involved
stakeholders. In this direction, new applications and services can be
implemented by enabling advanced monitoring functionalities in the main
ITS actors (e.g., car, buses, trucks). Considering the public transport scenario,
classical monitoring solutions adopted in buses consist of a reduced set
of data collected through the On-Board Diagnostic (OBD) system, and in
turn sent towards a control center by using Automatic Vehicle Monitoring
(AVM) devices acting as system gateways. Such monitoring services can be
easily expanded by leveraging on advanced technologies considered the
pillars of the so-called Internet of Things (IoT) revolution: wireless sensor
devices organized in monitoring networks. Wireless Sensor Networks
(WSNs) can be easily integrated in the bus environment to gather additional
data able to: (i) provide a better service to passengers, (ii) improve route
planning strategies, and (iii) increase reliability of buses through on-line
maintenance services.

All the above mentioned features are addressed by the OBIT (Open
Bus in the Internet of Things) system developed by the Italian National
Inter-university Consortium for Telecommunications (CNIT) and Tiemme

---

*Corresponding author: matteo.petracca@sssup.it

S.p.A., a public transport company operating in the southern part of Tuscany, in Italy. OBIT is an in-vehicle wireless monitoring solution based on IoT devices which is able to measure in real-time additional parameters inside the bus environment. The first part of the chapter presents the OBIT system by detailing its architecture, components and functionalities. In the second part the experience in the OBIT system deployment is reported before to show obtained experimental results in a real testbed. On the field, experiments are presented for two working days in order to prove the proper functioning of the developed solution. The last part of the chapter presents a cost analysis based on the deployment experience, while final remarks conclude the document.

1. Introduction

In recent years the use of Information and Communication Technologies (ICTs) in the transport sector has enabled the creation of advanced smart services for all modes of transport, thus enabling the creation of Intelligent Transport Systems (ITSs) effectively. Though the term ITS generally refers to all modes of transport, the European Union Directive 2010/40/EU [1] defines ITS as systems in which ICTs are applied in the field of road transport with the aim of improving safety, efficiency and sustainability.

The great interest in advanced ITS solutions has led to a global standardization process. At the international level the ISO/TC 204 is the workgroup responsible for the system and infrastructure aspects of ITSs, while at the European level the standardization effort is led by ETSI TC ITS with several working groups. The main goal of such a standardization process is to foster the development of global ITS applications, thus overtaking the limits of the first ITS installations based mainly on proprietary solutions. In such a standardization process new technologies have been proposed during the years and their use standardized in order to extend the reference ISO/ETSI architecture for ITS [2]. This is the case of the Wireless Sensor Networks (WSNs) technology based on the Internet of Things (IoT) protocol solutions, whose adoption in the ITS scenario has been recently envisioned [3] by proposing two new standards [4, 5].

WSNs in ITSs can be used both in road side [6, 7, 8, 9] and vehicular [10, 11] segments by extending already deployed monitoring services to create advanced smart applications in the envisioned Smart Cities. In this respect the chapter presents the OBIT (Open Bus in the Internet of Things) system, an advanced monitoring solution based on wireless sensor devices, specially developed for the public transport scenario with the aim of extending already existing in-vehicle monitoring systems, usually based on data collected through the On-Board Diagnostic (OBD) system. The remaining part of the chapter is organized as follows. The next section discusses the main ITS applications in the Local Public Transport (LPT) scenario by emphasizing the benefits of in-vehicle WSN solutions, then the OBIT design is presented as well as its implementation, deployment and experimental which results in a real testbed. A cost analysis based on the deployment experience is reported in the last part of the chapter before conclusion.

2. ITS Applications for the Local Public Transport

From a general point of view the whole set of ITS applications can be classified in three categories acting in three main directions [12]: (i) transportation safety, (ii) transportation efficiency, and (iii) user services typically devoted to connectivity and convenience. In the local public transport scenario ITSs are used mainly to support decision-making processes in respect of efficiency and end user satisfaction. For instance ITS solutions can be successfully used to:

- reduce the travel time of bus lines, and consequently the waiting time of people at bus stops;
- improve fleet and personnel management policies;
- increase the comfort and quality of the service to increase the attractiveness of public transport in end users;
- rebalance the relationship between LPT and private transport through the development of integrated multimodal mobility systems.

To support the above mentioned decision-making processes what is strictly necessary is:

- to geographically localize the whole fleet through the use of Automatic Vehicle Monitoring (AVM) systems able to communicate in real-time with a control room;
- to immediately inform passengers at bus stops (e.g., by means of information poles) and inside buses (e.g., by means of smartphones);
- to monitor the LPT actual usage by measuring the real number of passengers for each line during the whole day;
- to monitor the real fuel consumption, as well as the mechanical degradation of components which may cause a service interruption;
- to monitor passengers' comfort, while increasing the perceived security and safety levels.

While the use of AVM systems with adequate information policies is nowadays a very popular solution in the LPT scenario to increase passengers' satisfaction and to improve fleet and management policies, a further progress in providing a better service can be reached by adopting additional monitoring solutions. For instance, the detection of data such as the actual number of passengers, the fuel consumption, the generated pollution, etc., can be used to plan the whole local public transport service of a certain city better in order to be more effective and green with lower costs. At the same time additional monitoring solutions able to communicate additional parameters with respect to those already provided by the OBD system can be installed to plan the vehicles maintenance better, while environmental sensors can be used to improve the passengers' satisfaction.

Though the possible above mentioned monitoring solutions can be easily added in new buses, their use in already existing vehicles requires the use of technologies able to retrofit the bus environment. In this respect, WSNs can be easily integrated in the bus environment as in-vehicle solution interfaced with already available AVM systems with the aim of gathering additional data able to: (i) provide a better service to passengers, (ii) improve route planning strategies, and (iii) increase reliability of buses through on-line maintenance services. In fact WSNs are ideally based on battery powered devices and their deployment does not need the installation of additional cables. Moreover, by supporting standard protocols compliant with the IoT vision, each sensor device can be easily integrated in a standard in-vehicle monitoring system based on well-known protocols coming from the Internet world, thus supporting interoperability among systems. Each sensor device can just be considered as a source of simple [13] or complex [14, 15, 16] information that can be retrieved simply, or, on the contrary, it can also process the available information in order to take decisions according to a certain logic that can be changed during its life cycle by using over the air software reprogramming techniques [17] or virtual machine based approaches [18, 19, 20].

3. The OBIT System

In order to support the above mentioned decision-making processes through the use of additional in-vehicle monitoring solution, the OBIT system has been designed. OBIT is the first in-vehicle monitoring system in which the WSN technology is applied in the LPT domain to retrofit the bus environment. To support passengers' comfort better, to improve route planning and to increase reliability of buses the following categories of sensors have been selected:

- temperature sensors to detect whether the air conditioning system in the bus is working or not;
- fuel oil meter sensor to measure the instant fuel consumption;
- oil condition sensor to measure in any instant the oil degradation in order to prevent possible engine failures.

From a high-level point of view, the OBIT architecture has been designed for a seamless integration of new wireless sensors with already installed automatic vehicle monitoring systems, thus creating a real suitable solution to retrofit buses. The whole OBIT architecture is depicted in Fig. 1 where its main components are highlighted, and it is composed of:

(a) The wireless sensor network segment consisting of several Sensing Devices (SDs) with wireless communication capabilities. All SD nodes are organized in a monitoring network able to send data to a Border Network Element (BNE), connected with the AVM system, through standard protocols developed to enable IP-based communications in wireless sensor networks. The considered protocols are: (i) IEEE.802.15.4 at the physical and medium access control layers, (ii) 6LoWPAN (IPv6 over Low power Wireless Personal Area Networks) and RPL (IPv6 Routing Protocol for Low-Power and Lossy Networks) at the network layer, and (iii) CoAP (Constrained Application Protocol) at the application layer. An overview of such communication protocols can be found in [21], they represent the basis to build open systems compliant with the Internet of Things requirements;

(b) The AVM On-Board Computer (AVM-OBC), the in-vehicle AVM device able to gather data inside the vehicle, as well as to send them to a remote server by means of a remote connection. From a general point of view it is a device able to manage multiple network interfaces (system gateway) such as GPRS, 3G and WiFi. Moreover, it must be connected with the BNE node in order to gather data from the in-vehicle WSN. The connection between the AVM-OBC and the BNE can be done through a simple Ethernet connection in order to retrofit the system. In the future, the BNE could simply be another network interface really embedded into the AVM-OBC;

(c) The Control Room (CR), the last part of the system, in which all data coming from multiple AVM-OBCs are collected, analyzed and stored inside a central database. Moreover, from the CR possible feedbacks can be sent towards buses in real-time.

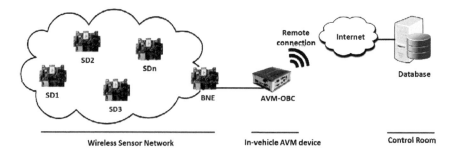

Figure 1: OBIT architecture.

In the developed system SD and BNE components are based on the Seed-Eye board, a microcontroller-based device developed as basic IoT platform able to host both simple (e.g., temperature, humidity, light) and complex (e.g., micro cameras) sensors. The Seed-Eye, fully described in [22] and depicted in Fig. 2a, has been selected because of its communication module, fully compliant with the IoT protocol stack, and its processing capabilities, able to support advanced sensing requirements. On such a device, temperature, fuel oil meter and oil condition sensors have been integrated to create the above introduced SD nodes.

For temperature monitoring purposes an analog sensor commercialized by Thermo King has been integrated on the Seed-Eye through a dedicated connection board, while an appropriate software application has been developed to improve the measurement accuracy and to convert measured electric levels in Celsius degrees (°C) as final output of the device. Although ideally each SD device should embed a single sensor, thus creating a single wireless sensor node, in the developed system two special SD devices have been created, embedding two and three temperature sensors respectively. The SD node embedding two sensors is reported in Fig. 2b. The choice of creating such special SD devices has been made because of the necessity of collecting the temperature in several points, while reducing at the same time the amount of devices to be installed inside the bus during the testbed deployment. The use of two temperature SD devices instead of five is a practical simplification only; at a higher level of view such a choice can be considered a device virtualization approach. Thanks to the five temperature sensors integrated on two Seed-Eye boards, the OBIT system is able to collect the internal bus temperature in four points, thus providing a much more accurate measure of the internal temperature, as well as the external temperature in a single point.

To create a sensing device able to measure the instant fuel consumption the CONTOIL® DFM8 EDM fuel oil meter by Aquametro [23] has been integrated in the Seed-Eye, an image of such SD device is reported in Fig. 2c. The selected fuel oil meter has been developed specially for fleet management purposes and it is suitable for differential-based fuel injection systems. Albeit in this case the hardware integration has been easily performed by using standard communication buses, the software integration has been demanding in terms of software development because of the necessity of managing real-time processing tasks able to successfully detect the sensor output. The validation of the developed software solution has been done through a real fuel consumption simulator developed by Tiemme S.p.A. in its garage. The conceptual schema of such a simulator is reported in Fig. 3a, while the real system is reported in Fig. 3b. The integrated SD device is able to measure the fuel consumption in a given window of time in terms of liters per hour (l/h).

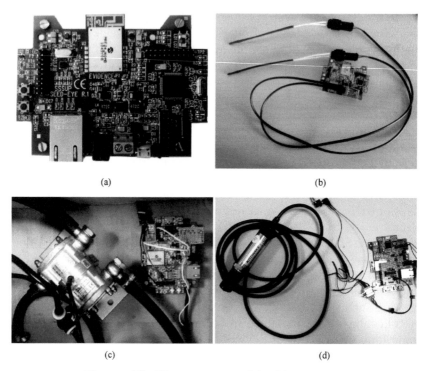

(a)   (b)

(c)   (d)

Figure 2: The SD components of the OBIT system.

The last monitoring component of the system is the SD node able to measure the motor oil degradation. To evaluate such degradation the sensor MA-K16136-KW by Kittiwake [24] has been selected as market solution to be integrated in the Seed-Eye board. A picture of this last SD node is reported in Fig. 2d. The sensor is able to provide in real-time a measure of the oil quality, as well as its temperature. The oil quality measure is provided in the percentage of degradation with respect to an initial condition evaluated through a calibration process, while the oil temperature is provided in Celsius degrees (°C). Because of the sensor interface, an additional hardware component has been designed for the physical integration, while a simple software based on command requests and interpreted answers has been developed.

In order to test the developed in-vehicle WSN solution without interfering with the already installed AVM systems a new AVM-OBC has been developed. Such a component is an effective solution able to gather data from the WSN while sending them towards the CR thanks to a GPRS connection. Moreover, the developed AVM-OBC embeds a GPS device to referencing all collected data in time and space, as well as a WiFi interface to show data inside the bus through smartphones. The AVM-OBC is

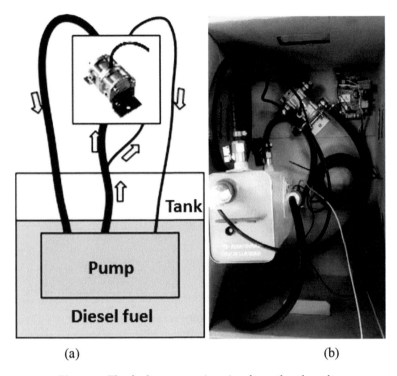

(a)   (b)

Figure 3: The fuel consumption simulator developed to
test the fuel consumption SD node.

mainly composed by the BeagleBone Black board [25], depicted in Fig.
4a, a low-cost device based on an ARM® Cortex-A8 microprocessor and
developed by CircuitCo in association with Texas Instruments. On top
of the BeagleBone Black board an expansion module enabling GPRS and
GPS functionalities is mounted, see Fig. 4b, while the WiFi connectivity
is provided by a USB dongle. The AVM-OBC is connected with the BNE
trough called the Ethernet interface and it is able to gather data from the
WSN for sending them through the GPRS connection, as well as store
them for a certain amount of time in order to show information through an
embedded website. A snapshot of the embedded website reporting data
gathered from all SD nodes is reported in Fig. 5. To store all information
in a specific database installed in the control room a proper software
solution has been developed. Such a software is mainly a web server able
to receive data from the AVM-OBC on one side and able to store them
in a MySQL database on the other side. Moreover, the software provides
several application programming interfaces for future extensions of the
system (e.g., connection with other software tools already installed for
management purposes).

(a)   (b)

Figure 4: BeagleBone Black without additional interfaces (left)
and with GPS/GPRS expansion (right).

BUS

Data   Settings   Admin   Logout

SN	Time	Latitude	NS	Longitude	EW	
4	May 29, 2014, 2:34 p.m.	0	0	0	0	Details
3	May 29, 2014, 2:34 p.m.	0	0	0	0	Details

Temperature sensors

2-T1	2-T2	2-T3	1-T1	1-T2
Internal	Internal	External	Internal	Internal
28.8 °C	29.1 °C	28.9 °C	27.3 °C	28.3 °C

Average Internal Temperature: 28.4 °C
Average External Temperature: 28.9 °C

Oil condition sensor

Quality: 1 Temperature: 31 °C

Fuel consumption sensor

Consumption: 35.0 l/h

cnit   fiemme

Figure 5: BeagleBone Black embedded web server showing
last acquired data from SD nodes.

It must be stressed again that although in the OBIT development a
new AVM-OBC device has been created, this has been done in order to test
the new added monitoring functionalities in isolation, without interfering
with already working AVM systems. To really retrofit buses with the new
WSN-based technology the BNE must be physically integrated with the
existing AVM-OBCs which must be able, from a software point of view,
to manage the new interface to read new information. Moreover, an
additional database table must be added in the AVM central software, thus
allowing the system to store data coming from SD nodes for a subsequent
analysis activity.

4. The OBIT System in a Real Bus

In this section of the chapter the real deployment of the OBIT system is presented to discuss experimental results obtained with the system in operation for several days.

4.1 OBIT Deployment

To experiment the OBIT functionalities in real conditions the whole system has been deployed on a selected bus of the Tiemme S.p.A. fleet, the 1720. Such a bus has been selected because it is already equipped with an AVM system, thus enabling the possibility of a posteriori correlation between data gathered by both systems to confirm the OBIT potential and benefits.

Though WSN devices are generally battery powered, in the system deployment all devices have been connected to a power source in order to focus the first experimentation on the functionalities of the whole system, thus without considering energy issues and leaving them as main topic to be considered in future works. To connect each OBIT device to a power source, an additional electrical wiring has been performed inside the bus. In order to decouple the original electrical system from OBIT components, a dedicated DC-DC converter has been installed inside the bus power control panel. Starting from such a component several electric cables have been installed inside the bus behind plastic panels following the schema reported in Fig. 6. In the picture the bus power control panel is labelled as AL.

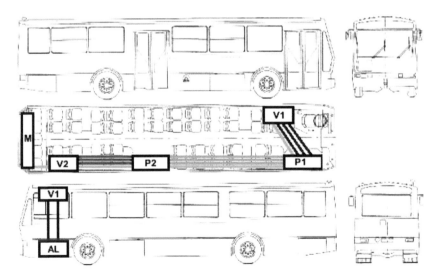

Figure 6: Electrical wiring schema for the power supply of the OBIT components.

To place SD nodes, as well as the developed AVM-OBC, two main compartments have been created (V1 and V2); moreover just on top of each bus entrance two additional compartments have been added for future developments of the system (i.e., to install visual sensors able to count the number of people inside the bus). In V1 the SD node with two temperature sensors and the AVM-OBC have been installed, Fig. 7a, while the SD nodes embedding three temperature sensors, the fuel oil meter and the mechanic oil condition devices have been installed in V2, Fig. 7b.

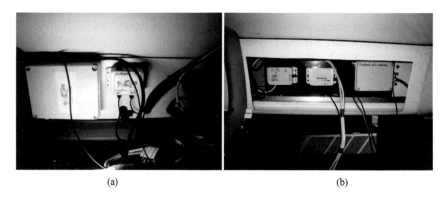

(a)   (b)

Figure 7: SD and AVM-OBC nodes installed inside the bus
(V1 on the left and V2 on the right).

To measure the internal temperature at several points, the temperature sensors integrated on the Seed-Eye board have been equipped with a longer cable with respect to the one reported in Fig. 2b, and the four internal sensors have been installed close to the driver location (Fig. 8a), in the middle of the bus both in the bottom and the top (Figs 8b and 8c), and in the rear (Fig. 8c). The external temperature sensor has been installed close to the front right light (Fig. 9).

The fuel oil meter and the mechanical oil condition devices connected to the Seed-Eye have been installed in the engine compartment by modifying the fuel-injection system (the CONTOIL® DFM8 EDM must be installed between the delivery and return fuel ducts), as well as the engine oil system (the MA-K16136-KW must be in an oil bath to work properly). A picture of both sensors inside the engine compartment is reported in Fig. 10, while they are shown separately in Figs 11a and 11b, where it is possible to notice the solutions developed to install the two components in the bus engine.

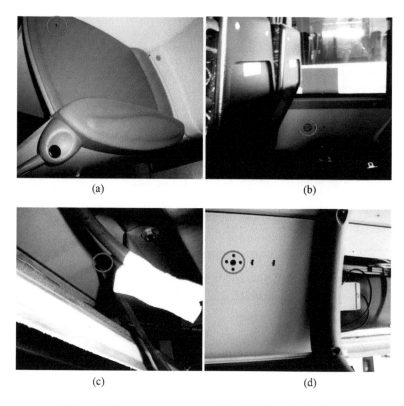

Figure 8: Temperature sensors installed inside the bus.

Figure 9: Temperature sensor installed outside the bus.

Figure 10: CONTOIL® DFM8 EDM and MA-K16136-KW in the engine compartment.

Figure 11: Installation details for CONTOIL® DFM8 EDM (left) and MA-K16136-KW (right).

4.2 Experimental Results

The proper functioning of the deployed system has been tested for several days by splitting the experimentation in two parts. In the first part all OBIT components have been tested, while in the second part, after the occurrence of a critical event, some components of the system have been removed. For both periods the data acquired for a whole bus work shift in a selected day are reported. In the following the label Day 1 and Day 2 will be used to indicate the two selected days. In all presented results the OBIT system has been configured to gather data every minute, therefore all graphs will report a per point based plot.

Experimental results obtained on Day 1 have been selected as an interesting example of OBIT behavior to be shown because they refer to the day on which the critical situation occurred. On such a day the whole bus work shift goes from 14:52 pm to 22:40 pm with several breaks according to what is reported in Table 1.

Table 1: Bus work shift during Day 1

ID	Start	End
TFS	14:52	15:00
KC 620	15:00	16:05
TFS	16:05	16:23
GR 550	17:00	17:55
GA 660	18:20	19:20
NR 706	20:30	20:58
NR 573	21:27	21:57
NA 580	22:10	22:40

The behavior of the internal and external temperature during the whole day is reported in Fig. 12. As it is possible to see, the average internal temperature is always higher than the external, even though they have the same behavior. This is because the bus was obviously under the sun, and since it was not a very hot day the air conditioning system was in off.

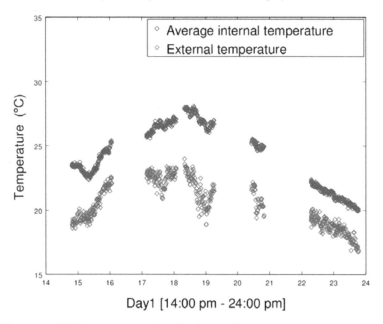

Figure 12: OBIT experimental results during Day 1: temperature behavior.

Results regarding the fuel consumption and the mechanical oil temperature are reported in Figs 13 and 14 respectively. In all presented results the mechanic oil degradation percentage is missing; in fact, although such data have been correctly acquired by OBIT, their values do not have a real sense since it was not possible to change the oil of the engine at the beginning of the experimentation. As it is possible to see from the reported graphs, the instant fuel consumption is greater than zero when the bus is circulating and at the same time the mechanical oil temperature increases.

The fuel consumption and mechanical oil temperature behaviors described above are not the same from 22:00 pm to 24:00 pm. In such a window of time the bus control panel was in an 'on' state with the engine off, and this is also the reason why the mechanical oil temperature is decreasing. In the last functioning window time a critical situation occurred: the additional tubes installed to modify the engine oil system broke down causing the breakdown of the engine. Although the encountered situation caused the suspension of the OBIT experimentation, this has led to a much more accurate analysis of the acquired data correlating them with those acquired by the AVM system already installed in the bus, thus showing that the developed OBIT system was able to detect the same condition with the same accuracy of a commercial AVM system. The correlation analysis has been conducted by analyzing as a function of the time the

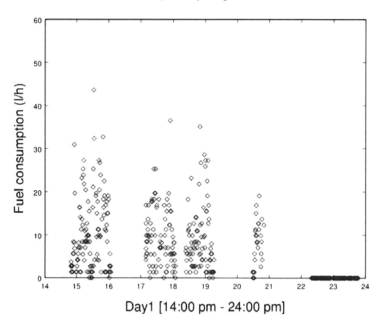

Figure 13: OBIT experimental results during Day 1: fuel consumption.

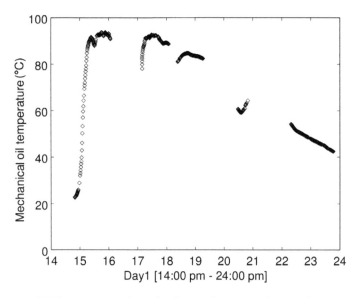

Figure 14: OBIT experimental results during Day 1: mechanic oil temperature.

OBIT fuel consumption sensor output and the AVM engine on/off signal. The condition in which the OBIT system is working sending a fuel consumption equal to zero appears at the same time in which the AVM system detects several engine on/off signals. Moreover, while several engine on/off signals suggest a possible problem inside the bus, the fuel consumption behavior gives an additional information, thus suggesting a possible mechanical problem.

The OBIT experimentation has continued after the restoration of the bus engine, even though with temperature sensors only. Day 2 results belong to such a second period and they refer to a work shift going from 8:20 am to 20:00 pm according to what is reported in Table 2.

Table 2: Bus work shift during Day 2

D	Start	End
BA 140	8:20	8:40
HR 060	12:15	13:15
FA 030	13:38	14:25
NR 205	14:28	14:48
NA 150	15:10	15:40
FR 511	16:05	16:25
HR 560	16:35	17:00
AT 260	17:30	18:30
FG 050	18:40	20:00

Such a specific day has been selected to show how OBIT is able to understand whether the air conditioning system is working properly, thus allowing the bus service provider to evaluate the passengers' comfort. The temperature behavior during Day 2 is reported in Fig. 15. As it happens on Day 1, during the first part of the day the internal temperature is higher than the external because the air conditioning system is not activated yet. In the middle of the day, after a starting point around 12:00 am, the internal temperature is lower than the external, thus showing that the air conditioning system is working properly. When the driver turns off the air conditioning system the internal temperature slowly increases and this is what happens from 18:40 pm to 20:00 pm.

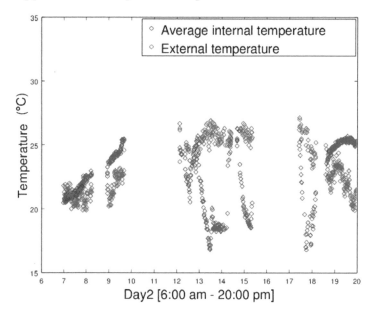

Figure 15: OBIT experimental results during Day 2: temperature behavior.

The OBIT experimentation on a real testbed is a first real deployment of a WSN system inside a bus. Such experimentation has shown on the one hand the feasibility and the benefits of an in-vehicle monitoring system able to retrofit the bus environment; on the other hand it has shown as during the retrofitting installation, particular attention must be given to crucial systems to be modified in order to avoid severe failures.

5. Cost Analysis

This section of the chapter provides an OBIT cost analysis mainly based on the cost of the hardware components used for the testbed. Since the installation has been performed in several runs, and its effort will be greatly reduced in a future version of the OBIT system based on battery powered devices, no costs related to the installation process have been considered and reported. The overall costs divided for hardware components are reported in Table 3, as it is possible to see that the total cost of the OBIT system is lower than 3500€, an amount of money that can be considered marginal with respect to the cost of a whole bus. Moreover, the overall OBIT costs can be further reduced by leveraging on economies of scale in case a whole bus fleet must be equipped, thus making the OBIT system much more attractive.

Table 3: Detailed costs for the OBIT hardware components

Hardware components	Costs without taxes
5 Thermo King temperature sensors	150€
1 CONTOIL® DFM8 EDM	735€
1 MA-K16136-KW	1240€
5 Seed-Eye boards	500€
1 BeagleBoneBlack with GPS/GPRS expansion, WiFi dongles and antennas	250€
1 DC-DC converter	45€
Various materials for the installation	400€
Total costs	3320€

Considering OBIT composed by three main monitoring systems able to gather data regarding (i) temperature, (ii) fuel consumption and (iii) mechanical oil condition, the cost for each system is graphically summarized in Fig. 16, and reported in analytical terms below:

- temperature monitoring system: 500€;
- fuel consumption monitoring system: 985€;
- mechanical oil condition monitoring system: 1585€
- AVM-OBC: 250€

The highest costs are given by mechanical oil condition and fuel consumption monitoring systems which have an impact of 48% and 30% respectively. Such costs are mainly due to the commercial sensors connected to the Seed-Eye, the MA-K16136-KW and the CONTOIL® DFM8 EDM. Anyway, thanks to the modularity and flexibility introduced by the adopted IoT protocol solutions each component of the OBIT system can be installed independently from the others, thus allowing the bus service provider to create customized solutions within its fleet to reduce the total costs. Moreover, once a bus is divested the OBIT components can be easily installed in another bus, thus recovering part of the investment.

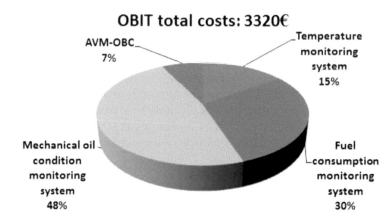

Figure 16: OBIT cost analysis as a function of monitoring systems.

6. Conclusions

In the chapter an in-vehicle monitoring solution based on IoT-based wireless sensor networks is presented for the local public transport scenario. The final system, OBIT (Open Bus in the Internet of Things), is able to retrofit the bus environment by allowing the additional monitoring of (i) in-bus temperature, real fuel consumption and mechanical oil condition with the aim of supporting decision-making processes in respect of better efficiency and give end user satisfaction. The data provided by the OBIT system can be successfully exploited to: (i) provide a better service to passengers, (ii) improve route planning strategies, and (iii) increase reliability of buses through on-line maintenance services.

In the chapter the OBIT system is first detailed in all its components, then experimental results obtained by deploying the system in a real bus have been presented for two selected working days. In the last part of the chapter, a cost analysis is presented showing that OBIT has a marginal cost with respect to the whole bus. Moreover, thanks to the modularity and flexibility introduced by the adopted IoT protocol solutions, each component of the OBIT system can be installed independently from the others, thus allowing the bus service provider to create customized solutions within its fleet. In this direction, a further flexibility can be achieved through AVM-OBC devices able to support multiple interfaces able to interact with wireless sensors compliant with different standards. To support several possible low-level IoT technologies, future improvements of the OBIT system, besides the use of battery powered sensor devices, will be based on the AMBER (Advanced Mother Board for Embedded systems pRototyping) board [26], an open-source device recently developed with the aim of supporting advanced applications in the IoT ecosystem fully.

Acknowledgements

The authors want to thank Tiemme S.p.A. in the figures of Piero Sassoli (General director), Massimiliano Pellegrini (Technical director), Maurizio Pelosi (Responsible of technological systems devoted to the customer service support), Marcello Pianigiani (Technical staff), and Raffaele Masselluci (Technical staff) for their full support in the OBIT development and experimentation.

References

1. [EU] European Parliament. 2010. Directive 2010/40/ EU on the framework for the deployment of intelligent transport systems in the field of road transport and for interfaces with other modes of transport. *Official Journal of European Union*. Lyon, France.
2. [ISO21217] International Organization for Standardization. 2014. Communications access for land mobile (CALM) – Architecture. ISO21217:2014 Intelligent Transport Systems.
3. Pagano, P., M. Petracca, D. Alessandrelli and C. Salvadori. 2013. Is ICT mature for an EU-wide intelligent transport system? Intelligent Transport Systems, IET, vol. 7, no. 1, pp. 151–159.
4. [ISO19079] International Organization for Standardization. 2015. Communications access for land mobile (CALM) – 6LoWPAN networking. ISO19079:2015 Intelligent Transport Systems.
5. [ISO19080] International Organization for Standardization. 2015. Communications access for land mobile (CALM) – CoAP facility. ISO19080:2015 Intelligent Transport Systems.
6. Alessandrelli, D., A. Azzarà, M. Petracca, C. Nastasi and P. Pagano. 2012. ScanTraffic: Smart Camera Network for Traffic Information Collection, European Conference on Wireless Sensor Networks, pp. 196–211.
7. Salvadori, C., M. Petracca, M. Ghibaudi and P. Pagano. 2012. On-board Image Processing in Wireless Multimedia Sensor Networks: A Parking Space Monitoring Solution for Intelligent Transportation Systems. pp. 245–266. In: *Intelligent Sensor Networks: Across Sensing, Signal Processing, and Machine Learning*. CRC Press. Taylor & Francis Group, London, New York.
8. Salvadori, C., M. Petracca, S. Bocchino, R. Pelliccia and P. Pagano. 2014. A low-cost vehicle counter for next-generation ITS. *Journal of Real-Time Image Processing*. Springer Berlin, Heidelberg. pp. 1–17.
9. Carignani, M., S. Ferrini, M. Petracca, M. Falcitelli and P. Pagano. 2015. A Prototype Bridge Between Automotive and the IoT. IEEE World Forum on Internet of Things, pp. 1–6.
10. Pellerano, G., M. Falcitelli, M. Petracca and P. Pagano. 2013. 6LoWPAN conform ITS-Station for non safety-critical services and applications. International Conference on ITS Telecommunications, pp. 426–432.
11. Bellè, A., M. Falcitelli, M. Petracca and P. Pagano. 2013. Development of IEEE802.15.7 based ITS services using low-cost embedded systems. IEEE International conference on ITS Telecommunications, pp. 419–425.

12. Zhang, T. and L. Delgrossi. 2012. *Vehicle safety applications: Protocols, security and privacy.* Wiley Press, Hoboken, New Jersey.

13. Talzi, I., A. Hasler, S. Gruber and C. Tschudin. 2007. PermaSense: Investigating Permafrost with a WSN in the Swiss Alps. Fourth Workshop on Embedded Networked Sensors, pp. 8–12.

14. Salvadori, C., M. Petracca, S. Madeo, S. Bocchino and P. Pagano. 2013. Video streaming applications in wireless camera networks: A change detection based approach targeted to 6LoWPAN. *Journal of Systems Architecture*, pp. 859–869.

15. Petracca, M., G. Litovsky, A. Rinotti, M. Tacca, J.C. De Martin and A. Fumagalli. 2009. Perceptual based voice multi-hop transmission over wireless sensor networks. IEEE Symposium on Computers and Communications, pp. 19–24.

16. Petracca, M., C. Salvadori, S. Bocchino and P. Pagano. 2014. Error Resilient Video Streaming with BCH Code Protection in Wireless Sensor Networks, *Journal of Communications Software and Systems*, vol. 10, no. 1, pp. 31–40.

17. Mottola, L. and G.P. Picco. 2011. Programming Wireless Sensor Networks: Fundamental Concepts and State of the Art. *ACM Computing Surveys*, vol. 43, no. 3, pp. 1–51.

18. Alessandrelli, D., M. Petracca and P. Pagano. 2013. T-Res: Enabling reconfigurable in-network processing in IoT-based WSNs. IEEE International Conference on Distributed Computing in Sensor Systems, pp. 337–344.

19. Azzarà, A., M. Petracca and P. Pagano. 2015. The ICSI M2M Middleware for IoT-based Intelligent Transportation Systems. IEEE International Conference on Intelligent Transportation Systems, pp. 155–160.

20. Bocchino, S., S. Fedor and M. Petracca. 2015. PyFUNS: A Python Framework for Ubiquitous Networking Sensors. European Conference on Wireless Sensor Networks, pp. 1–18.

21. Azzarà, A., S. Bocchino, P. Pagano, G. Pellerano and M. Petracca. 2013. Middleware solutions in WSN: The IoT oriented approach in the ICSI project. IEEE International Conference on Software, Telecommunications and Computer, pp. 1–6.

22. Petracca, M., S. Bocchino, A. Azzarà, R. Pelliccia, M. Ghibaudi and P. Pagano. 2013. WSN and RFID Integration in the IoT scenario: An Advanced Safety System for Industrial Plants. *Journal of Communications Software and Systems*, vol. 9, no. 1, pp. 104–113.

23. [AQUAMETRO] CONTOIL® VZD/VZP and DFM. Online at: http://www.aquametro.com/data/docs/en/2934/aquametro-4500d-contoil-dfm.pdf. Last access on April 22nd.

24. [KITTIWAKE] Oil Condition sensor. Online at: http://www.kittiwake.com/sites/default/files/Oil%20Condition%20Sensor%20Brochure%20MA-K19077-KW.pdf. Last access on April 22nd.

25. [BEAGLEBONE BLACK] BeagleBone Black project. Online at: http://beagleboard.org/black. Last access on April 22nd.

26. [AMBER] AMBER project. Online at: http://amber-lab.com. Last access on April 22nd.

Data Management and Data Sharing in Field Operational Tests

Yvonne Barnard[1*], Sami Koskinen[2], Satu Innamaa[2], Helena Gellerman[3], Erik Svanberg[3], Adrian Zlocki[4] and Haibo Chen[1]

[1]Institute for Transport Studies, University of Leeds, UK
[2]VTT Technical Research Centre of Finland Ltd, Finland
[3]SAFER Vehicle and Traffic Safety Centre at Chalmers University of Technology, Sweden
[4]Institut für Kraftfahrzeuge (IKA) at RWTH Aachen University, Germany

Abstract

In this chapter it will be discussed how data from Field Operational Tests of Intelligent Transport Systems can be managed and shared. The Field Operational Tests, where hundreds of users get to experience the latest systems, aim to assess the impacts that would result from a wide-scale implementation. Evaluation principles of Field Operational Tests will be explained, and a closer look will be taken at the data that is collected for carrying out the assessments. The widely used FESTA methodology for designing and conducting Field Operational Tests and Naturalistic Driving Studies already provides several recommendations for managing data. This methodology will be discussed and illustrated by examples of its use in European projects. As field test projects set out to collect a huge set of data, the projects themselves do not usually have the scope or the resources to analyze the data from every perspective. Therefore re-use of the collected data also by other projects with different research questions has the potential to generate a wealth of new knowledge about what is happening in the interactions between drivers, vehicles and the infrastructure. Data sharing is the focus of a European support action, FOT-Net Data. The support action is working, with international collaboration, to form a data sharing framework, a data catalogue, and provide detailed recommendations for

---
*Corresponding author: Y.Barnard@leeds.ac.uk

sharing and re-use. Outcomes from this activity will be discussed. Ways of sharing different types of data will be described, including the necessary steps to be taken to open up the data.

1. Introduction

Over the past decades new technologies for Intelligent Transport Systems (ITS) have been researched and developed. In Europe these activities have often been supported by the European Research Framework Programmes as well as national programmes. The short and long term impacts of these systems need to be understood to answer questions which are crucial for market introduction and penetration. By testing the systems on a large scale, in real driving conditions during a significant period of time, Field Operational Tests (FOT) can answer most of these questions. The results of FOTs enable policymakers to establish the right framework for deployment of these systems, and business leaders to make informed decisions about their market introduction. Since 2008 the European Union has supported a number of projects enabling testing of the latest vehicle information technology in large-scale field trials. Thousands of drivers have been able to test the most promising prototypes or products just entering the markets. The drivers have been testing systems such as adaptive cruise control, forward collision warning, navigators and most recently, warning systems based on short-range wireless communication between vehicles. The communication can provide information on, for example, nearby car accidents or approaching emergency vehicles. Field test projects have evaluated the impact of these technologies, and contributed to their introduction. Drivers' behaviour whilst using the systems has been monitored for continuous periods of up to more than a year, collecting valuable information from traffic.

In the European Commission funded FESTA project a common FOT methodology was developed [1, 2], which is now being widely used as the basis for the planning and execution of FOTs and Naturalistic Driving Studies (NDS). The naturalistic studies concentrate on studying everyday driving behaviour and events that may lead to accidents. Since the original FESTA project, the methodology has been maintained and updated by European FOT-Net [3, 4] support actions.

The FOTs have a crucial need for a platform of knowledge exchange in order to let individual FOTs benefit from each other's experiences. Without such exchange, it takes long especially for newcomers to learn the several practical steps of arranging field trials. A platform was set-up in 2008 as the FOT-Net support action, funded by the EC DG Information Society. The first FOT-Net ended in 2010 and was followed by FOT-Net 2 from 2011 to 2014, and FOT-Net Data, which will end in 2017. FOT-Net supports and coordinates the FOT community by sharing experiences, providing methodology training, building a knowledge base of past tests, maintaining and updating the FESTA methodology and promoting the FOT results. FOT-Net has also developed strategic networking for national, European and global FOTs (mainly North American and Asian-Pacific). The latest support action in the series, FOT-Net Data, promotes especially the sharing and re-using of data gathered in FOTs. Its main objectives are in development of a global data sharing framework in collaboration with world-wide stakeholders and keeping a catalogue of available datasets. All information and documents concerning the FOT-Net support actions are available at the FOT-Net website [3].

The Data Gathered in FOTs and the Knowledge Derived from the Analysis

FOTs are performed to gather knowledge about the impact the large-scale introduction of ITS may have. The socio-economic impacts that are usually considered are:

- Safety: will ITS improve road-safety, for drivers but also for other road-users?
- Mobility: will ITS have a direct effect on traffic flow and mobility, and indirect effects such as reduced crashes and will it change mobility patterns of citizens?
- Environmental: will ITS reduce CO_2 emissions and pollutants, for example, by changed driving behaviours or reduced congestion?
- Economical: what will ITS mean for the costs of systems in vehicles and on the road, and for societal costs such as those caused by congestion?

If one looks closer into how these impact areas are influenced one can use the six areas of impact defined by FESTA [1] based on Draskóczy et al. [5]. Although this approach was originally designed for formulating hypotheses on traffic safety impacts, it is in fact equally well applicable for efficiency and environmental impacts.

The six types of effect are [1, 6]:

1. Direct effects of a system on the user and on driving
2. Indirect (behavioural adaptation) effects of the system on the user
3. Indirect (behavioural adaptation) effects of the system on non-users (imitating effect)
4. Modification of interaction between users and non-users (including vulnerable road users)
5. Modification of accident consequences (e.g. by improving rescue, etc. — note that this can affect efficiency and environment as well as safety)
6. Effects of combination with other systems

In FOTs a variety of complex data is gathered, relative to the specific research questions. Three types of data may be distinguished.

Data collected by sensors in a vehicle, for example, vehicle controls such as speed or turn indicators, or position of steering wheel or pedals, GPS traces, vehicle kinematics from accelerometers, or video recorded from one or many cameras mounted in the vehicle, pointing to the driver or e.g. the front view. Such data was, for example, collected in the first European large-scale FOT euroFOT [7]. Depending on the different vehicles (from different manufacturers) within the field tests different objective data was collected from vehicle dynamics sensors, environment perception sensors or additional image processing data from cameras filming the scenario, the driver and his interaction with the control elements of the vehicle. Objective vehicle data was processed and evaluated for the research questions of the project [8].

Data collected by road-side units and back-end systems, including contextual data, such as traffic flow and weather data, and communication data. For example, the DRIVE C2X project stored messages communicated between cars and road-side units, and traffic flow information derived from inductive loops in the road infrastructure. Internet map and weather data sources were also used in enriching vehicle data, enabling assessment of, e.g., speeding behaviour in different weather conditions [9].

Subjective data gathered from the participants in the FOT, such as questionnaire data and travel diaries. For example, the TeleFOT project collected subjective data of participants' uptake and acceptance with four consecutive questionnaires (before field tests started, twice during the tests and after the tests were completed) [10]. Another source of subjective data in TeleFOT was travel diaries. Participants were asked to fill in a travel diary (timing, length, mode, origin, destination and purpose of travel) for one week similarly four times during the FOT with the same frequency as the questionnaires to collect data on potential mobility impacts [11].

In addition to the collected raw data, FOT datasets often consist of derived data and aggregated data. Derived data usually consists of cleaned-up, filtered and resampled versions of the raw signals. Aggregated data represents computations of different aspects of data segments. It may be the average speed during a trip or the number of passes through a specific intersection.

Another dimension of the data is subjective manual video annotations, often generated when reviewing specific events in retrospect. Real-time observations can also help to provide useful understanding of FOT data.

The analysis of data gathered by FOTs can provide evidence concerning the effects of ITS introduction, supporting the decision making of industry and policy makers. It also provides a variety of data on what is actually happening on the roads, and how drivers behave, knowledge that will be very useful for future transport research.

2. Collecting and Managing Data in FOTs

As FOTs are complex studies, a structured approach is needed to set-up the study, collect and analyze the data and determine the impact of the systems evaluated. In 2008 the FESTA methodology was developed to provide guidance and support for the upcoming EU-funded FOTs. A handbook was produced with many practical recommendations [1]. The basis of this handbook was a methodology, to be followed by the FOTs. This methodology has not only been adopted by FOTs funded by the European Commission but also by many nationally (or otherwise) funded projects, and it has influenced FOTs outside Europe. The FESTA methodology is summarized in Fig. 1. It includes several steps, which, although described in linear way, are performed in iteration. The V-shape shows the dependencies between the different steps on the left- and right-hand side of the V. The main steps can be summarized as follows:

- Defining the study: Defining functions, use cases, research questions and hypotheses
- Preparing the study: Determining performance indicators, study design, measures and sensors
- Conducting the study: Collecting data
- Analyzing the data: Storing the data, analyzing the data, testing hypotheses, answering research questions
- Determining the impact: Impact assessment and deployment scenarios, socio-economic cost-benefits analysis

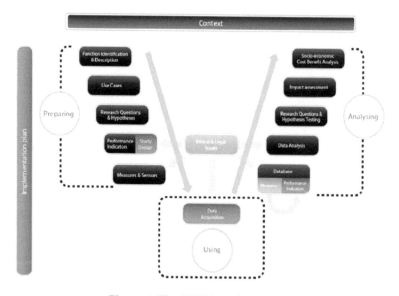

Figure 1: The FESTA methodology.

FOTs produce large amounts of complex data and the main challenge is to make the data manageable and easily available for analysis [12]. As each FOT has its own type of data and data management demands, a generic model that is suitable for all FOTs does not exist. Often the data recorders lay the foundations of the FOT dataset by grouping data into segments, e.g. by trips or events. By establishing these segments it is possible to attach rich metadata that can help filter the dataset.

Documentation is a key requirement. The data need to be described in detail, but also the tools and processes implemented during the FOT. To understand the dataset the study design (the objectives of data collection) is important information, as well as the detailed test protocols with relevant test scenarios. It is also important to describe the structure of the dataset, how it is organized and stored, to facilitate use of the data.

Data quality assurance is another key issue. Procedures ensuring the quality have to be applied from the very beginning of the planning and piloting of the data collection, and the subsequent storage and processing.

3. FOT Data Collection: Projects and Data Catalogues

FOT projects gather large data-sets, usually from the three sources described above: in-vehicle sensors, road-side units (for projects studying cooperative systems) and from drivers. This section shortly describes several large FOTs and gives examples of the data types they have collected, but this selection is by no means exhaustive. In the FOT-Net wiki, a catalogue of FOTs may be found, listing over 150 projects [4].

The largest datasets have so far been collected in the US (e.g. IVBSS [13], SHRP2 [14] and Safety Pilot [15]) and in Europe (e.g. euroFOT [19, 20], TeleFOT [21], DriveC2X [16], FOTsis [17], and UDRIVE [18]).

The IVBSS project designed four Integrated Vehicle-Based Safety Systems, and tested the systems in cars and heavy trucks [13]. Data includes vehicle sensor signals, video data, questionnaires and focus groups. The world´s largest naturalistic driving study SHRP2 includes data from normal driving from over 3000 participants during 1-2 years each (SHRP2) [14]. Data includes in-vehicle signals, internal and external videos, radar, questionnaires, weather data and detailed map data. The Safety Pilot was a connected vehicles project where over 2000 vehicles are instrumented and the communication data is collected [15]. On some vehicles, data about the driver behaviour is also collected (GPS, speed, vehicle signals, video data, questionnaires).

In Europe several large-scale FOTs [4] have collected data (some of the projects are listed above). The euroFOT project was one of the first large-scale FOTs of in-vehicle Advanced Driver Assistance Systems (ADAS) performed on the roads of several European countries [19, 20]. Data collection included vehicle dynamics, data from environment perception sensors and video. The TeleFOT project studied the use of nomadic devices. Data collection included GPS data, video data, travel diaries, questionnaires, focus groups and interviews [21]. FOTsis [17, 22] and DRIVE C2X [16, 23] studied cooperative systems. Data collection included communication data, GPS and HMI data, video data, questionnaires, focus groups and interviews. UDRIVE [18] is the first large European naturalistic driving study on cars, trucks and scooters. Data collected includes video data from eight cameras (including a smart camera), GPS data, vehicle dynamics data and a range of participant questionnaires (UDRIVE) [18, 24].

In Japan large data sets based on event recorders have been collected, such as the Tokyo University of Agriculture and Technology (TUAT) event recorder dataset (GPS, speed, accelerometer, brake pedal, turn indicator, and forward video), and Australia as well has several interesting datasets, e.g. the Australian NDS [25] collecting naturalistic data from the normal driving of 360 participants (video, GPS, accelerometer, radar, questionnaires) [26, 25]. Data collection also takes place in Korea and China.

In most of these studies, equipped vehicles were used by ordinary volunteer drivers for over a long period (months or even years) in real world traffic conditions and large amounts of data have been gathered by sensors, capturing driver and vehicle behaviour, as well as the driving context. The data has been used to answer a wide range of research questions. These data enabled the analysis of the impacts of ADAS and cooperative systems on safety, efficiency and the environment, as well as the usability and acceptance of the systems by the users [27, 28, 21, 29]. The data is also used to analyze crash causation [30, 31]. The size of the datasets varies, from gigabytes (GB) to several petabytes (PB), depending mainly on whether the data is collected continuously, and whether it includes video.

Although datasets may be huge, and contain valuable information, often not all data collected is analyzed within a project, either due to lack of time or resources, or because more data is collected than is needed to answer the selected and specific research questions. This data can be of value for research purposes other than those of the project that collected them, but this requires an awareness of its existence, the willingness and capability of its owner to share it.

In Europe datasets reside with the project or the partners in consortia that gathered the data. FOT datasets are not stored in public repositories. The FOT-Net Data support action [3, 4] (see below) is currently working on a central catalogue, where researchers looking for FOT data can find what data is available, and how they may obtain it.

The closest counterpart to the European FOT catalogue activities is the Research Data Exchange [32], which is a core part of USDOT's Data Capture and Management Program. The RDE is a tool to access transport research data, especially on connected vehicle technologies. The data shared on RDE is anonymized, partly with the help of dedicated software tools.

While the FOT-Net wiki [4] already includes a comprehensive catalogue of field trials and naturalistic driving studies carried out in recent years across the world, the FOT-Net Data support action is compiling further details regarding available research datasets and tools related to them. The catalogue is moderated by FOT-Net [3, 4], but as a wiki it continues to receive input and updates from the FOT community – various organizations carrying out large-scale trials.

The main purpose of the FOT Data Catalogue is to support potential data re-users in identifying suitable datasets for their purposes, and to facilitate data sharing. The main principle of the catalogue is that the actual datasets remain with their owners. Data providers will make the final agreements with interested organizations, and can offer support to new analysts regarding the details of the study. The catalogue will include information on data and contacts but not the data itself. It is, however, possible to add anonymized sample data for allowing re-users to get a practical example.

This division of content between the FOT Data Catalogue and data owners is presented in Fig. 2. The figure also lists common roles: catalogue

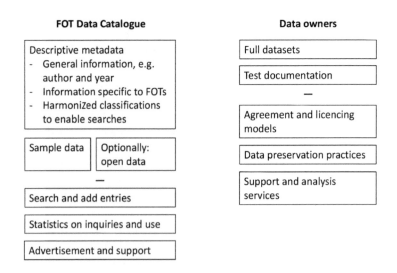

Figure 2: FOT Data Catalogue relationship with data owners.

operators collect statistics of information searches, and can advertise the datasets available from a community. Catalogue operators also offer basic support to researchers, while the data owners and persons who executed the test are the ones who have in-depth knowledge of it.

4. Data Sharing: Benefits, Challenges and Recommendations

Data sharing means that not only the research community, but also industry and public authorities, will have access to a wealth of information. Data from one project may be of value to other projects. For example, most FOTs collect data about the speed of vehicles. This is not only of interest to a project investigating the effect of speed adaptation systems on safety, but may also be of interest to researchers and stakeholders interested in very different questions, for example, the speed patterns of different age groups, eco-driving behaviour, etc. Re-using data for further and new analyses seems to be the obvious answer, as collecting new data, in a new project, is very costly.

Sharing and re-using data can be a major step forward in our understanding of the behaviour of transport users and systems. It will be very hard to find sponsors for large numbers of FOTs that start from scratch with their own data collection. In Europe the European Commission is stressing this need for data to be open and shared, as they are one of the main sponsors of data collection. A wealth of information is hidden in the datasets that have been collected in recent years, so the question is how one will be able to generate new knowledge out of these existing datasets.

Opening datasets allows other institutes to verify impacts and calculations. This is an important part of scientific processes and publications. Access to data also allows for calculations from larger populations, assessing the situation in several countries. Data sharing should also be seen as a method of cooperation: the group that originally collected data obtaining complementary results from new researchers. These results will often be interesting for the original funders, and may boost new projects and consortia.

However, sharing poses challenges, which may be of a technical, legal and organizational or practical and financial nature [33].

Technical challenges concern issues such as the quality of metadata, descriptions of implementations, how field tests were run, how the data was collected, the tools used to collect and store the data and the standards and formats used.

Legal and organizational issues concern ownership, data protection and privacy issues. For example, permission from the FOT participants is needed to allow third parties access to the data. Re-engineering tested services or used sensor systems are also seen as problematic.

Practical and financial issues concern questions about who is paying for the access, are there resources to be allocated to support data sharing, the training of new data analysts so they can understand the data, its limitations, tools, and the physical access to the data?

And finally, there are issues such as trust and willingness to share that may depend on earlier experiences and on how the challenges mentioned above are met. In order to re-use data in a new project, the original project must be willing to share the dataset and make it available in a way ensuring that it can be re-used. The implementation of the solutions to the challenges that is chosen is therefore crucial.

To support FOT organizers and data providers in addressing these issues, the FOT-Net Data support action provides a framework including guidelines and recommendations [34, 35]. The Data Sharing Framework consists of seven different areas, which should all be addressed in order to provide or re-use data.

1. Agreements within the project collecting data, including consortium agreements, participant agreements and agreements with third party data providers
2. Availability of valid data and metadata, including a "standard" description of the documentation of the data
3. Data protection requirements both for the data provider and the analysis site
4. Security and personal integrity education for all personnel involved
5. Support and research services to facilitate the start-up of projects and offer research capabilities
6. Financial models to provide funding for the data to be maintained and available and for access provision personnel to be available
7. Application procedures and data sharing agreements

Each chapter gives hands-on advice on what to implement, and includes checklists and topics to discuss both in the original project and in the data sharing agreement. The availability of a common Data Sharing Framework will importantly facilitate a larger use of the collected FOT/NDS data. The organizations setting up new FOT/NDS projects will not need to develop the specific data sharing content for a certain project, but will be able to focus on the project specific questions such as research questions, study design, and data acquisition requirements. Also, researchers wanting to re-use already collected datasets will be able to make use of a more or less standard application procedure, rely on already performed training that is widely accepted, and plan for the costs that using a specific dataset may imply for the project.

5. Conclusions: The Future of Open and Shared Data

While FOTs [4] mostly focussed on testing systems implemented in vehicles and/or infrastructure, the next large step will be the field testing of automated driving, including driverless vehicles. This means that there is an even stronger need for data sharing. As fundamental changes in traffic will start to take place, all kinds of insight in the impacts are needed to support public authorities, industry and the general public in forming their opinions and strategies. Only if data are shared widely will it be possible to find evidence on the effects of the introduction and use of automated vehicles. Industrial competition may be a main obstacle, but it is in everyone's interest to ensure that the introduction of automation does not have negative consequences on impact areas such as safety and traffic efficiency.

Not only the sharing of new data is of interest for automation, also the existing FOT data are very useful. Data on relevant driving situations will be necessary for the design and development of automated systems. Also other evaluation methods, such as virtual simulations, need this data. This "relevant situations" data cannot be collected by one company or consortium only, so sharing is a necessity.

Data sharing and openness have long been wished for. The benefits are clear and numerous, but there are many difficult steps to be taken. Guidelines, agreements, and even changes of opinion are needed. The strongest force for effecting these changes is the setting of new conditions for public funding. When data is gathered with public money, certain levels of openness are required.

In the ITS field, during the last fifteen years, public transport data and timetables have increasingly become available via open internet interfaces required by funding organizations. Additionally, spatial map data and city databases are opening up. These allow for large data analyses to be carried out. Emerging new research topics will benefit from new analyses of existing data.

Datasets from single ITS research projects will remain diverse and will require detailed documentation to be analyzed. These datasets can, however, be increasingly open, and can be collected according to established practices. Standardization of certain key parts (position, map, communication, etc.) of data may be foreseen, and may make analysis and documentation of data more straightforward. However, up to the present moment even GPS data has always been logged in slightly different formats. The varying formats require flexibility from analysis software, requiring first setting up reader components to gather the information required for analysis.

FOT-Net Data [3] and similar efforts will define the minimum level of documentation and data to collect in FOTs, and set up public data catalogues. These being available, some major hurdles, such as absence of information on what is available, of detailed documentation of the contents of a dataset, of the conditions for its re-use, and even on whether or not the participants of the study have given their consent for re-use, are no longer in the way.

When projects can refer to existing documentation, the effort of data collection is likely to be reduced, instead of placing further burden on them. Harmonized processes and methodology also contribute to the collection of high-quality data that is of real interest for further work. Finally, re-use of datasets is gradually gaining ground in scientific work, and it should no longer be a principle that all scientists gather their own dataset to work with.

Acknowledgements

FOT-Net Data is a Coordination and Support Action in the seventh Framework Programme Information and Communication Technologies. It is funded by the European Commission (EC), DG Connect, Grant Agreement number 610453.

References

1. FESTA. 2014. FESTA Handbook, Version 5. Available at: http://wiki.fot net. eu/index.php?title=FESTA_handbook (accessed on 28.4.2016).
2. Barnard, Y. and O. Carsten. 2010. Field Operational Tests: challenges and methods. In: J. Krems, T. Petzoldt, T. and M. Henning (eds.), Proceedings of European Conference on Human Centred Design for Intelligent Transport Systems, pp. 323–332. Lyon: Humanist Publications. Lyon, France.
3. FOT-Net. http://fot-net.eu (accessed on 28.4.2016).
4. FOT-Net wiki. http://wiki.fot-net.eu (accessed on 28.4.2016).
5. Draskóczy, M., O. Carsten and R. Kulmala. 1998. Road safety guidelines. Deliverable B5.2 of CODE project (TR1103). Atkins Wootton Jeffreys, Birmingham, UK.
6. Carsten, O. and Y. Barnard. 2010. Preparing field operational tests for driver support systems: a research-oriented approach. In: D. de Waard, A. Axelsson, M. Berglund, B. Peters and C. Weikert (eds.), *Human Factors: A system view of human, technology and organisation*. Maastricht, the Netherlands: Shaker Publishing. pp. 243–248.
7. Benmimoun, M., A. Zlocki, C. Kessler, A. Etemad, W. Hagleitner, M.A. Obojski and W. Schwertberger. 2011. Execution of a field operational test within the euroFOT project at the German test site. In: Proceedings of the 18th World Congress on Intelligent Transport Systems, Orlando.
8. Fahrenkrog, F., M. Benmimoun, A. Zlocki and L. Eckstein. 2011. Analysis of radar sensor information for an impact assessment of driver assistance

systems in the European field operation test "euroFOT". 8th International Workshop on Intelligent Transportation (WIT 2011), Hamburg.

9. Koskinen, S., E. Svanberg, R. Rotstein, J. Rech, F. De Ponte Müller, F. Häusler and H. Wedemeijer. 2013. Report on implementation of DRIVE C2X data management to test sites, DRIVE C2X Deliverable D35.1. Available at: http://www.drive-c2x.eu/publications

10. Karlsson, M., M. Alonso, R. Brignolo, S. Innamaa, A. May, A. Meriga, O.M. Perez, A. Rämä and T. Skoglund. 2013. Impact on user uptake – results and implications, TeleFOT Deliverable D4.7.3.

11. Innamaa, S., M. Axelson-Fisk, L. Borgarello, R. Brignolo, L. Guidotti, O. Martin Perez, A. Morris, K. Paglé, A. Rämä, P. Wallgren and D. Will. 2013. Impacts on Mobility – Results and Implications, TeleFOT Deliverable D4.4.3.

12. Koskinen, S. 2013. Data Management in Field Operational Tests. In: Proceedings of the 22nd ITS World Congress 2013 in Tokyo.

13. IVBSS. http://www.its.dot.gov/ivbss (accessed on 28.4.2016).

14. SHRP2. http://www.trb.org/StrategicHighwayResearchProgram2SHRP2/Blank2.aspx (accessed on 28.4.2016).

15. Safety Pilot. http://safetypilot.umtri.umich.edu (accessed on 28.4.2016).

16. DRIVE C2X. http://www.drive-c2x.eu/project (accessed on 28.4.2016).

17. FOTsis. http://www.fotsis.com

18. UDRIVE. http://udrive.eu (accessed on 28.4.2016).

19. Benmimoun, A., M. Benmimoun, I. Wilmink and M. van Noort. 2009. euroFOT: Large Scale Field Operational Tests – Impact Assessment. In: Proceedings of the 16th ITS World Congress, Stockholm, Sweden.

20. Benmimoun, A., M. Benmimoun, M. Regan, M. van Noort and I. Wilmink. 2010. Challenges in analysing data from a field operational test of advanced driver assistance systems: the euroFOT experience. In: Proceedings of Transport Research Arena 2010, Brussels.

21. Mononen, P., S. Franzen, K. Pagle, A. Morris, S. Innamaa, M. Karlsson, K. Touliou, R. Montanari and S. Fruttaldo. 2013. Final Report, TeleFOT Deliverable D1.15.

22. Alfonso, J., N. Sánchez, J.M. Menendez and E. Cacheiro. 2013. An ITS architecture specification – The FOTsis project experience. Securitas Vialis, pp. 61-77.

23. Schulze, M., T. Mäkinen, T. Kessel, S. Metzner and H. Stoyanov. 2014. Final Report, DRIVE C2X Deliverable D11.6. Available at: http://www.drive-c2x.eu/publications (accessed on 28.4.2016).

24. Eenink, R., Y. Barnard, M. Baumann, X. Augros and F. Utesch. 2014. UDRIVE: The European naturalistic driving study. In: Proceedings of the Transport Research Arena, Paris.

25. ANDS. http://www.tars.unsw.edu.au/research/Current/Australian_Naturalistic_Driving_Study/australian_nds.html (accessed on 28.4.2016).

26. Regan, M.A., A.Williamson, R. Grzebieta, J. Charlton, M. Lenne, B. Watson, N. Haworth, A. Rakotonirainy, J. Woolley, R. Anderson, T. Senserrick and K. Young. 2013. The Australian 400-car Naturalistic Driving Study: Innovation in road safety research and policy. In: Proceedings of the 2013 Australasian Road Safety Research, Policing & Education Conference, Brisbane, Queensland. Australasian Council of Road Safety.

27. Barnard, Y., F. Fischer and M. Flament. 2015. Field Operational Tests and Deployment Plans. In: C. Campolo, A. Molinaro and R. Scopigno (eds.), *Vehicular ad hoc Networks: Standards, Solutions, and Research*. Springer, pp. 393–408. Cham, Switzerland.

28. Malone, K., J. Hogema, S. Innamaa, S. Hausberger, M. Dippold, M. van Noort, E. de Feijter, P. Rämä, E. Aittoniemi, T. Benz, H. Enigk, I. Giosan, C. Gotschol, D. Gustafsson, I. Heinig, K. Katsaros, D. Neef, L. Ojeda, R. Schindhelm, C. Sütterlin and F. Visintainer. 2014. Impact assessment and user perception of cooperative systems, DRIVE C2X Deliverable D11.4. Available at: http://www.drive-c2x.eu/publications

29. Malta, L., M. Ljung Aust, F. Freek, B. Metz, G. Saint Pierre, M. Benmimoun et al. 2012. Final results: Impacts on traffic safety. euroFOT Deliverable D6.4.

30. Victor, T., M. Dozza, J. Bärgman, C. Boda, J. Engström, C. Flannagan and Schäfer, R. 2014. Analysis of Naturalistic Driving Study Data: Safer Glances, Driver Inattention, and Crash Risk SHRP 2 safety project SO8A. Transportation Research Board of the National Academies. SHRP, vol. 2.

31. Bärgman, J., V. Lisovskaja, T. Victor, C. Flannagan and M. Dozza. 2015. How does glance behavior influence crash and injury risk? A 'what-if' counterfactual simulation using crashes and near-crashes from SHRP2. Transportation Research Part F: Traffic Psychology and Behaviour, vol. 35, pp. 152–169.

32. RDE. http://www.its-rde.net (accessed on 28.4.2016).

33. Barnard, Y., S. Koskinen and H. Gellerman. 2014. A platform for sharing data from field operational tests. In: Proceedings of the 21st ITS World Congress 2014 in Detroit.

34. Gellerman, H., E. Svanberg, J. Bärgman and Y. Barnard. 2015. Data Sharing Framework for Naturalistic Driving Data. In: Proceedings of the 24th World Congress on Intelligent Transport Systems in Bordeaux.

35. Gellerman, H., E. Svanberg, I. Heinig, C. Val, S. Koskinen, S. Innamaa and A. Zlocki. 2015. Data Sharing Framework. Public Draft FOT-Net Data Deliverable D3.1. Available at: http://fot-net.eu/Documents/data-sharing-framework/ (accessed on 28.4.2016).

Design, Implementation and Field Trial of DSRC-based Transit Signal Priority System

Andy Jeng*

Industrial Technology Research Institute
195, Sec. 4, Chung Hsing Road, Chutung, Hsinchu,
Taiwan 31040, Republic of China

Abstract

Transit Signal Priority (TSP) is a relevant case of Intelligent Transportation System which can facilitate the movement of buses across signalized intersections but depends heavily on reliable communication technology. In this chapter, a WAVE/DSRC-based bus TSP system deployed at Hsinchu, Taiwan is introduced. To verify the benefits and possible impact, over 120 thousands of field trial records were collected from the test site over six months, and a series of analyses were conducted. The results show that the developed system can significantly reduce the intersection passing time and stopped times and bring very little impact to existing signal control system.

1. Introduction

Transit Signal Priority (TSP) is one of the important Intelligent Transportation Systems (ITS) [1, 2] that uses a combination of sensing and communication technologies to improve the passenger's experience and reduce urban congestion by giving transit vehicles higher priorities to pass through intersections. A bus TSP system is aimed at facilitating the movement of buses across signalized intersections through temporary traffic signal timing adjustment.

---

*Corresponding author: andyjeng@itri.org.tw

Compared to traditional signal control, a bus TSP system would be more sophisticated and complex. It should be able to detect approaching buses at a certain distance away in order to provide sufficient time to adjust the signals and grant priority while minimizing impacts on traffic, and may also require the ability to update dynamic information, e.g. the speed of the bus that can vary due to a number of traffic conditions. Therefore, the technology to detect approaching transits and collect real-time traffic information from them becomes critical.

A variety of technologies can be used as the candidates for bus detection, such as ground loop, Infrared, GPRS with GPS, Wi-Fi signal tower and Zigbee sensors. However, most of them were not specifically designed for vehicular environments, resulting in lower detection rates or requiring a higher cost to achieve the desired quality of service [3].

Wireless Access in Vehicular Environments and Dedicated Short Range Communication (WAVE/DSRC) technology specified in IEEE 802.11p/1609 [4, 5] is an amendment to the IEEE 802.11 standard. It enhances the support to ITS applications, with vehicle-to-vehicle (V2V) and vehicles-to-roadside (V2R) communications in the licensed 5.9 GHz band. In Europe, 802.11p is used as a basis for the ITS-G5 which is standardized by ETSI ITS.

Industrial Technology Research Institute (ITRI) is one of pioneers in supplying WAVE/DSRC qualified Communication Unit [6]. Recently, Taiwan's first WAVE/DSRC-based Bus TSP test site located at Hsinchu County was deployed. The region accommodates tens of thousands of residents from Hsinchu Science Park (HSP), one of Asian economic tigers. As shown in Fig. 1, the Bus TSP System is developed in collaboration with the county government and traffic signal controller supplier in

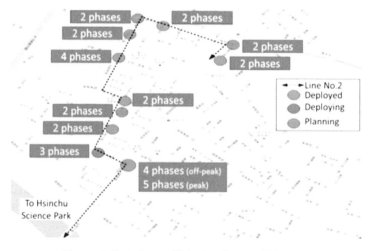

Figure 1: Test sites at Hsinchu County, Taiwan.

order to improve the efficiency of local express buses. The test site has been deployed on a 2-phase intersection along bus line No. 2 for proof of concept. Another test site of a more complex 3-phase intersection is undergoing.

This chapter introduces the WAVE/DSRC-based bus TSP system deployed at Hsinchu, Taiwan. Two real-time signal control strategies were implemented. The design issues as deployed in a real-world intersection is also discussed. To verify the benefits and impact, over 120 thousands of field trial records were collected from the test site over six months and a series of analyses were conducted. The results show that the system can save a third of passing time when buses pass through the intersection and can avoid being stopped more than half the times.

2. Field Trial System

Figure 2 shows the system architecture of the WAVE/DSRC-based Bus TSP operated in Hsinchu County, Taiwan. The entire system is composed of four major parts: (1) the On-board Unit (OBU), (2) the Roadside Unit (RSU), (3) the Traffic Signal Controller, and (4) the back-end management platform.

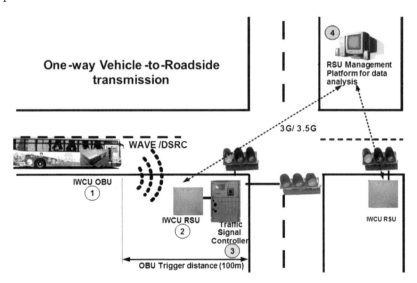

Figure 2: Architecture of ITRI's bus transit signal priority system.

As shown in Fig. 3 (a) and (b), each bus is installed an OBU (IWCU OBE 3.0) connected with an external antenna (5dBi gain) and power source. The OBU boosts automatically each time the bus starts and broadcasts a periodic beacon each second. The beacon format contains the current position and time info of the bus (from a built-in GPS receiver) in accordance with SAE J2735 DSRC Message Set Dictionary [7]. A RSU (IWCU RSE) is mounted on a traffic light pole (about 28 feet above the ground) at the intersection and connected with the traffic signal controller through an Ethernet port. Each IWCU RSE is also connected to a back-end IWCU RSU management platform via a 3G/3.5G cellar module for reporting error and post-data analysis.

The steps involved in providing priority are as follows: Each equipped bus will periodically send a message to announce its current position and related information. As shown in Fig. 3 (c), once the bus is detected at a pre-specified distance (100 m) upstream of the signal stop line by the RSE, a Priority Request Generator in the RSE will be notified of the approaching bus and alerts the traffic controller that the bus would like to receive priority. The controller processes the request and decides whether to grant priority based on defined conditions. The traffic controller then initiates action to provide priority to the bus based on the defined priority control strategies (see next section).

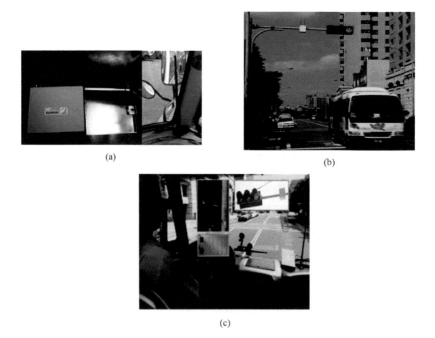

Figure 3: Installation: (a) OBU; (b) RSU; (c) Roadside Equipment.

2.1 Real-time Signal Control Logics

A variety of signal control strategies have been developed and implemented to adjust normal signal operation in favour of public transit and emergency vehicles, including Green Extension, Early Green, Queue-jump Phase, Phase Recall, Phase Suppression, and Phase Rotation [5]. The test site supports the first two modes, which have been implemented wildly in several bus TSP systems including both DSRC and non-DSRC based systems.

As shown in Fig. 4, green extensions are granted when a bus is expected to arrive a few seconds after the end of the green. In other words, if the intersection signals are already at a green phase for the approaching

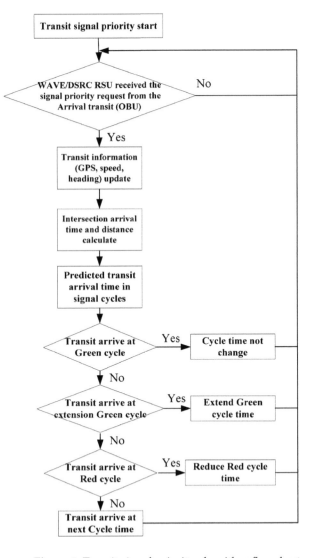

Figure 4: Transit signal priority algorithm flowchart.

bus and its distance is within 100 m, the controller will extend the length of the green phase to enable the bus to pass through the intersection on that phase. To achieve this, the extended phase time should be no less than the arrival time which can be estimated based on the bus's distance and average approaching speed.

On the other hand, early green recalls are granted to accommodate buses that would arrive a few seconds before the start of the green. In other words, if the intersection signals are at a red phase on the bus approach and the distance is within 100 m, the controller will shorten the green phase on the cross street to provide an earlier green phase for the bus approach to reduce the potential wait at the intersection. Note that in order to avoid risky situations to pedestrians on the cross street the shortening process should be postponed until the current green phase exceeds a certain period of time. One set 5 seconds at the test site.

For both cases, at the most one adjustment is allowed during each cycle. This is to avoid immense or unexpected influence on the original phase allocation. When the bus passes through the intersection, clearance is detected by the bus detection system and a communication is sent from the OBE to the traffic controller via the RSE that the bus has cleared the intersection. On being notified that the bus has cleared the intersection, the controller restores the normal signal timing through a predetermined logic.

3. Performance Evaluation

The field trial results of the system are now presented. As shown in Fig. 5, the RSE receives the beacons from OBE when it is in the transmission range and may send a request to the signal controller as long as the bus entered the pre-defined 100 m range. All records were automatically sent back to the backend RSE management platform for analytical purpose. Over the operation of the past six months 120,537 records have been collected and 1,427 signal control requests were successfully executed. The following measurements will be evaluated to verify the system:

1. Passing time: Average time between the bus entering the control area and leaving the stop line (i.e. points A and B in Fig. 5).
2. Travel time: Average time between the bus entering the prior and the later 100 m of the stop line (i.e. points A and C in Fig. 5).
3. Stopping times: Number of times the bus stopped (speed lower than 1 km/h over 3 s) before leaving point B in Fig. 5.
4. Speed distribution: Average speed at each distance between points A and C.

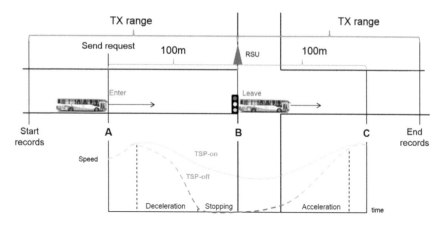

Figure 5: Measurements at an intersection.

Figures 6 and 7 show the passing time over different days in a week and different hours in a day. The passing with and without the bus TSP system (TSP-on vs. TSP-off) is compared. It can be seen that around 10 seconds can be saved before the bus passes through the intersection with the aid from prioritized signal control. The improvement is more significant during rush hours. As shown in Fig. 7, the passing time has the highest reduction during 6:00 ~ 8:00 and 16:00 ~ 18:00 (commuting time at HSP).

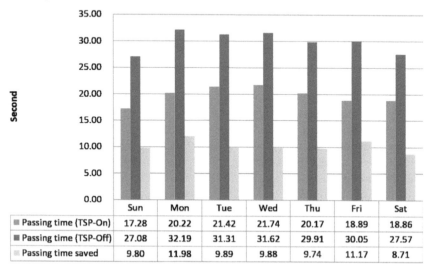

	Sun	Mon	Tue	Wed	Thu	Fri	Sat
■ Passing time (TSP-On)	17.28	20.22	21.42	21.74	20.17	18.89	18.86
■ Passing time (TSP-Off)	27.08	32.19	31.31	31.62	29.91	30.05	27.57
Passing time saved	9.80	11.98	9.89	9.88	9.74	11.17	8.71

Figure 6: Passing time by different days in a week.

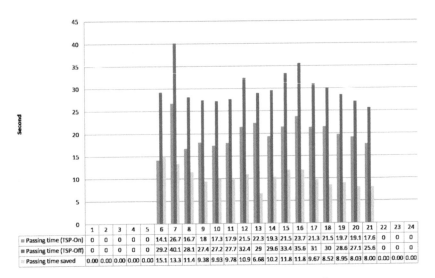

	1	2	3	4	5	6	7	8	9	10	11	12	13	14	15	16	17	18	19	20	21	22	23	24
■ Passing time (TSP-On)	0	0	0	0	0	14.1	26.7	16.7	18	17.3	17.9	21.5	22.3	19.3	21.5	23.7	21.3	21.5	19.7	19.1	17.6	0	0	0
■ Passing time (TSP-Off)	0	0	0	0	0	29.2	40.1	28.1	27.4	27.2	27.7	32.4	29	29.6	33.4	35.6	31	30	28.6	27.1	25.6	0	0	0
Passing time saved	0.00	0.00	0.00	0.00	0.00	15.1	13.3	11.4	9.38	9.93	9.78	10.9	6.68	10.2	11.8	11.8	9.67	8.52	8.95	8.03	8.00	0.00	0.00	0.00

Figure 7: Passing time by different hours in a day.

Figure 8 shows the performance gains in terms of reduced percentages of passing time, travel time and stopping times over different days and hours. It reveals that 33.76% of passing time can be reduced and 17.26% of travel time saved. The first measurement has a greater improvement, because the majority of time spent at an intersection is usually the waiting time caused by red phase before leaving the stop line. Interestingly, Fig. 8 (c) shows a surprisingly 51.62% of stopping times is avoided. It means that the system has a greater interest in improving passengers' satisfaction since frequent go-and-stop driving may easily cause the passenger to become impatient and thus unwilling to take buses. Besides, fuel consumption can be saved if bus drivers can avoid frequent breaking.

Taking a closer look at how the system makes improvements, Fig. 9 shows the distribution of the speed of the bus over different distances to the stop line. Each data point is averaged from all data records at the distance (round to meter). One can see that the bus TSP system can not only reduce the waiting time at the front of stop line but also avoid deceleration and thus keep buses at a comparatively constant speed.

Finally, it is analyzed how a priority control system impacts the traffic signal controller. As shown in Fig. 10, the TSP system has very a little side-effect to the original phase time allocation. It only increases 3.81% of green phase and reduces 3.31% red phase on an average (note that there are several sub states, called steps, in each phase. One omits the detailed discussion here). In other words, the system is quite suitable to be widely and independently deployed with almost no concern to the existing traffic control planning.

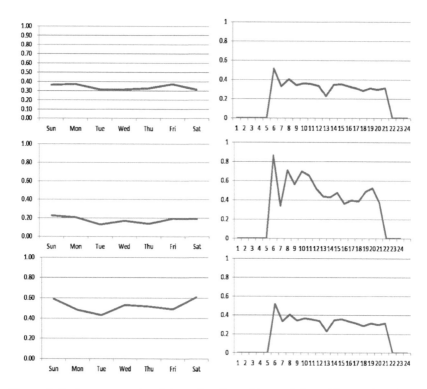

Figure 8: Performance gains: (a) passing time; (b) travel time; (c) stopping times.

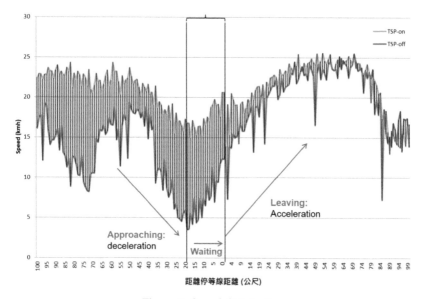

Figure 9: Speed distribution.

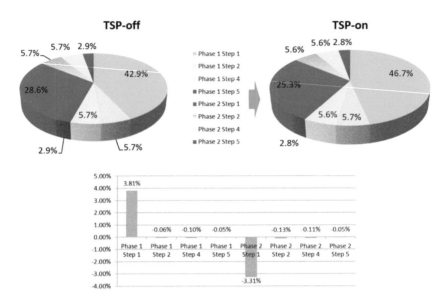

Figure 10: Impact to phase time allocation.

4. Conclusions

In this chapter, the WAVE/DSRC-based bus transit signal priority system deployed at Hsinchu, Taiwan, has been introduced. The green extensions and early green strategies have been implemented for a two-phase intersection. The field trial results collected over six months have been analysed and it has been shown that the system can significantly improve the passing time, travel time and stopped times, and have very little impact on the existing system.

The test site is continuing to serve while more and more passengers experienced its convenience. A 3-phase intersection along the same bus lines and performing a region-wide analysis to examine the benefits is being worked on if more intersections were engaged. One expects to present the results in future publications.

References

1. Tan, C.W., S. Park, H. Liu, Q. Xu and L.P. 2008. Prediction of transit vehicle arrival time for signal priority control: algorithm and performance, *IEEE Transaction on Intelligent Transportation Systems*, vol. 9, no. 4, pp. 688–692.
2. Zlatkovic, M., P.T. Martin and L. Tasic. 2011. Implementation of transit signal priority and predictive priority strategies in ASC/3 software-in-the-loop simulation. *Proc. of 14th Intelligent Transportation Systems Conference*, pp. 2130–2135.

3. Smith, Harrlet R., Brendon Hemily and Miomir Ivanovic. 2005. Transit Signal Priority (TSP): A Planning and Implementation Handbook. ITS America, U.S. DOT.
4. IEEE. 2010. 802.11p, Part II: Wireless LAN Medium Access Control (MAC) and the Physical Layer (PHY) Specifications, Amendment 6: Wireless Access in Vehicular Environment.
5. IEEE Standard for Wireless Access in Vehicular Environments (WAVE) – Networking Services. 2010. IEEE Std. 1609.3.
6. Li, H.H. and K.C. Lin. 2010. ITRI WAVE/DSRC Communication Unit. IEEE Vehicular Technology Conference (VTC-Spring), 2010, pp. 1–2.
7. SAE J2735. 2009. Dedicated short range communication (DSRC) message set dictionary.

Toward Smart Autonomous Cars

François Marmoiton* and Morgan Slade

Institut Pascal, UMR 6602 du CNRS, Campus Universitaire des Cézeaux
4 Avenue Blaise Pascal, TSA 60026, CS 60026
63178 Aubiere Cedex, France

Abstract

End of 2014, Institut Pascal and the Ligier Group presented an autonomous shuttle during the Michelin Challenge Bibendum (Chengdu, China). Like a horizontal lift, the vehicle can be called via terminals spread along its route. Users can then be automatically transported to their destination.

The purpose of this chapter is to describe, through the laboratory spectrum, the scientific and technical adventure that has created an object of this kind. Since the first concepts described by the pioneers in the 1980s, technical developments and scientific problem-solving have helped to imagine what this new mode of transport might be. The latest contributions aim to produce a driverless vehicle that is acceptable and reliable.

Firstly, it is presented how research laboratories tried to make smart cars in Europe in the late 1980s, while on-board electronic systems were in their infancy.

Secondly, the technical and scientific developments from 1995 to 2002 are presented. In fact, with the improvement of technology in the world of imaging, DGPS and processors, software solutions in laboratories made it possible to develop more accurate and reliable image processing outcomes and control of vehicles.

Thirdly, the building of the driverless shuttle EZ10 is presented step by step. During this period, the base of the guidance system was created and changes were made following the various experiments performed.

In perspective, the integration of the vehicle into its environment needs to be improved. The concepts to be developed are how to make it more acceptable and reliable. This requires increased interaction with the

*Corresponding author: francois.marmoiton@univ-bpclermont.fr

environment to make it a fully connected object, whether between vehicles of the fleet with users of the infrastructure or residents to the system.

Introduction

In 1900, steam-powered automobiles could be found alongside those fitted with electric motors or internal combustion engines. At that time, the means of propulsion was the major obstacle to the development of the automobile. The automobile began with a steam-powered engine (Cugnot cart in 1795), beat speed records thanks to electricity (the "Jamais Contente" was the first vehicle to break the 100 kph speed barrier in 1899) before mostly using the internal combustion engine later. The automobile was initially aimed at a wealthy clientèle with a certain spirit of adventure.

Once the technical problems were resolved, automobiles began to be produced industrially (in 1909 the Ford T was the first automobile to be mass produced) and then became an everyday consumer product. During the 20th century, the automobile underwent a series of technical or aesthetic developments. The First World War boosted mass production. Arms factories were effectively converted into car manufacturing plants. The Second World War boosted technological development with, for example, the appearance of the first automatic gearboxes. This development continued throughout the 1960s, which saw the appearance of the first premises of primary safety systems (Cyclone prototype fitted with two radar-sensing devices in 1959).

This momentum was halted by the 1970 oil crisis. After nearly a century of development, the car had become comfortable, high-performance and reliable, but was then perceived as polluting, energy-hungry, dangerous and impractical in town.

During the 1970s, draconian measures were taken to reduce energy consumption and improve passenger safety (speed limits, seat belts in front). The development of technologies and the miniaturization of systems in the electronics and computer fields led to the development of more complex systems. For instance, in the early 1980s, one saw the appearance of catalytic converters for exhaust pipes combined with electronic injection to reduce pollution or even primary (ABS, anti-wheel lock) and secondary (airbags, seat belt pretensioners) safety systems. In the following years, the performance of processors led to the creation of on-board systems that warn or assist the driver in very precise circumstances. That is why currently systems such as aids for reversing into a parking space, adaptive speed controllers or even road sign recognition systems being marketed are seen.

In laboratories, the first driverless cars started to make an appearance as early as the end of the 1980s. In the following years, the basic functions comprising localization and guidance systems for intelligent cars were defined and structured.

In this chapter, in first part, the pioneers' work on the intelligent systems built into cars are presented. This initial work was done over a period that began in the early 1980s and continued up to the mid-1990s.

Secondly, it will be shown how technological advances, in the second half of the 1990s, made it possible to imagine even more effective systems. Indeed, during this period, digital technologies increased the quality and stability of the information supplied by the on-board detectors. Benefiting from these new technologies and new algorithms for accurate guidance or localization, more precise, reliable vehicles started to take shape.

Thirdly, the performance of the guidance and localization systems made it possible to consider developing marketable products. One will then describe how a new mode of public transport, namely driverless shuttles, was devised and produced, from technologies developed at the start of this century.

Part I

1. Introduction

In the first half of the 1980s, the DARPA "Autonomous Land Vehicle" (ALV) demonstrator project was launched in the USA. At the same time, Moravec [1] in the USA, Tsugawa in Japan [2], and Ernst Dickmanns [3] in Europe were working on the problem of autonomous outdoor vehicles. Dickmanns then created the VAMORS vehicle, a small Mercedes van equipped with detectors, processors and actuators. This showed that it was possible to guide a vehicle using cameras (two cameras – one wide field, one narrow field) and a processor (10 Intel 8086) by following the white lines on a motorway or the edge of the traffic lane. As early as 1987, VAMORS drove at up to 96 kph automatically on a motorway under construction over a distance of 20 km.

Against this backdrop, as part of the European EUREKA research studies, the Prometheus programme (Programme for a European Traffic of Highest Efficiency and Unprecedented Safety) was set up as early as 1987.

Launched on the initiative of European car manufacturers and governments, Prometheus aimed to develop a technical base to encourage the development of road transport. So the objective was to come up with concepts and solutions to make road traffic safer, cheaper and with less impact on the environment, which would also make the road traffic system as a whole more efficient.

"Prometheus intends to fully exploit the potential of information and communication technology, micro-electronics, sensors and actuators and to adapt progresses in technology to the capabilities of human beings in order to create an optimum interaction between man and machine in road traffic." (F. Panik, 2007, The Prometheus Vision, ERTICO – ITS General Assembly and Partner Session on ITS Success Stories.

The project covers seven functional sub-systems:

- *Pro-Net:* Development of a communication network between vehicle-computers, enabling the vehicle to be driven safely by means of electronic sight, increasing the perception area of the driver,
- *Pro-Road:* Development of roadside communication and information systems to assist the driver and/or the onboard-computer to enable higher level traffic management functions,
- *Pro-Car:* Development of systems that interact with people like:
  - o computer-assisted systems in vehicles to assist and relieve the driver
  - o concepts to support safe interactions between driver and computer-assisted systems
  - o reliable, safe and high-quality Hardware and Software and concepts for diagnosis, service and maintenance,
- *Pro-Chip:* Development of the integrated microelectronics required for a Prometheus vehicle with advanced on-board automotive electronics,
- *Pro-Art:* Methodological investigations of applications of Artificial Intelligence for signal processing and decision-making processes,
- *Pro-Com:* Studies on the standardization of Vehicle to Vehicle (V2V) and Vehicle to Infrastructure (V2I) communication technologies,
- *Pro-Gen:* Research studies on the security on-board information systems.

Eleven automotive companies, 103 supplier companies and 124 research institutes will participate in the project, which encompasses all the technical aspects to make vehicles safer, from electronic components to artificial intelligence algorithms, not to mention the connection with the outside world.

In Germany, France and Italy, the research is organized around the national car manufacturers. In the electronics laboratory (now Institut Pascal, UMR6602 of the CNRS), in October 1988, one project was retained by the Prometheus office and formalized in a contract by the PSA group. The project concerns the study and implementation of a system to detect mobile obstacles ahead on the motorway through a combined 2D vision/laser range finding approach. At the time the team leader thought that the project "commits the laboratory for at least three years and should continue into the next century".

The laboratory [4] then structured its studies around three themes:

- parallel architectures to solve the problems tied to the constraints of video computation in real time (then 25Hz);
- the design and implementation of integrated intelligent sensors;
- the laboratory's third theme consisted in understanding and developing the necessary theoretical models for interpreting and analyzing images and sequences of images. The studies used statistical methods to make estimates of the movement and to try and estimate the actual movements of the detector and/or any objects of the scene observed.

Within the framework of the Prometheus project, the laboratory had to show the feasibility, in real-world conditions, of detecting mobile obstacles ahead on a motorway through an original approach, which consisted of merging the data from a 2D-type camera detector and a range-finder. The 2D camera data had to locate 3D-type objects by their perspective projection in the image shot, the range-finder would then provide further data on the distance, in the sight directions deduced from the image analysis.

The laboratory's main theme [5] was then on-board mobile vision (Vision Mobile Embarquée or VME in French) which it considered promising for top-level problems and industrial applications (assembly robotics; mobile cart in a known and semi-unknown world, autonomous vehicle in natural surroundings).

2. The First Achievements

The electronics laboratory had a positioning as integrator and experimenter capable of solving in advance the complex problems raised by the perception and interpretation of scenes from natural environments. The application field described at the time as "Active Dynamic Vision" directly concerned autonomous vehicles and notably the scientific obstacle of exteroceptive perception of the environment.

The objective of these studies in the long term was to produce sufficiently reliable visual perception systems to be integrated into a control process of a vehicle operating outdoors.

So this work relied on the teams' skills to develop four key functions:

- detect and track road obstacles,
- interpret road sequences,
- guide vehicles,
- implement on dedicated machines.

The spread of skills on each of these areas made it possible to master all the basic functions necessary to then create autonomous vehicles, namely detect, process and act.

So the EUREKA-Prometheus project created a real symbiosis between these studies in order to produce an electronic copilot capable of assisting the driver on the road. This project was followed by a CEMAGREF (IRSTEA now) project aimed at developing a service robot for natural areas, an automatic mower.

2.1 Detect and Track Road Obstacles

In this area, a new detector was designed [6] in order to successfully detect and track obstacles and particularly other vehicles on roads. This multiple-sensory detector combined a CCD video camera with a laser range-finder equipped with a scanning system (Fig. 1).

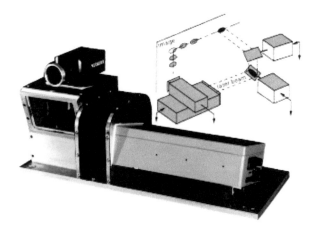

Figure 1: The multiple sensory detector.

Two operating strategies were developed for the detector in order to detect obstacles on the road:

- using information taken from images. After calibration of two detectors producing matches between the 3D range-finder impacts and the pixel coordinates of the image obtained by the camera, the presence of obstacles was obtained from the image data. The presence of obstacles was confirmed by the laser range finding in the windows of interest extracted from the luminance image.
- by following the reverse of the first strategy: the obstacles on the road are first detected only on the basis of depth data coming from the 3D detector. The obstacle detection and recognition became simpler, faster and more effective. This method required using the detector with a relatively low resolution (10 lines of 80 impacts) in order to meet the requirements in real time (200 ms with the detector of the day). The presence of vehicles was then confirmed by the vision data.

Once the vehicles were detected and recognized, a tracking stage estimates the position and speed relative to the obstacles more precisely. The tracking stage notably filters the observational noise of the detector with the aid of temporal filters (Fig. 2).

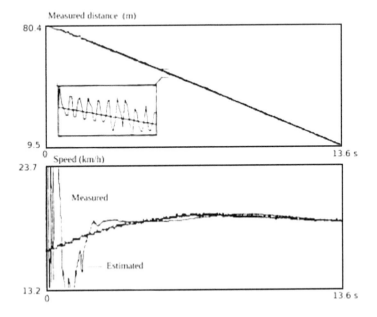

Figure 2: Position and speed estimations.

This algorithm was tested under real-world traffic conditions, on test tracks to detect the other vehicles using the Prolab 1 instrumented demonstrator (Fig. 3).

Figure 3: Prolab 1 instrumented demonstrator.

2.2 Interpret Road Sequences

In this field, the studies dealt with the knowledge and tracking of the motorway marking and detections of traffic lanes and service roads.

The measurement of the lateral positioning of the vehicle in comparison with the motorway white lines made it possible to conceive of driving aid applications, for instance, to alert the driver leaving a traffic lane or produce autonomous steering of the vehicle [7]. The algorithm used for this measurement relied on predictor-check techniques. A model of white lines of known geometry, because standardized, projected in the image, predicted the search areas. The detections were then made in these areas and used to update the model's three dynamic parameters (position of the vehicle on the road, angle of trajectory and curve of the road). No modelling pitch/roll phenomena can make the dynamic parameters estimate less accurate. The algorithm was sufficiently fast, easy to integrate and the 3D data obtained was precise enough to check a vehicle's position in relation to the road. Up until the early 2000s, further work was done to improve the reliability and precision of tracking methods and research strategies for primitive sights [8].

A study was done as part of Prometheus, in collaboration with Daimler-Benz, on the detection and recognition of road signs. This recognition was based on a segmentation by contours with a view to recognizing primitive shapes (circle, arc, triangle, etc.) that represent the basic components of the model to describe the road signs.

2.3 Vehicles Control

Using the algorithm that tracks the white lines described earlier, a vehicle can be automatically piloted from the visual information obtained by a monochrome camera located at the level of the rearview mirror. After modelling and kinematics and dynamic simulation of the vehicle, it was possible to establish the steering control rules to comfortably provide trajectory tracking at speeds in the region of 100 kph.

In collaboration with INRIA Sophia Antipolis and PSA, it was possible to test the dynamic control of a Citroën XM-type vehicle [9]. Different tests were carried out, notably like the bayonet test (two very tight successive bends) comfortably achieved up to a speed of about 100 kph.

2.4 Implement on Dedicated Machines

In the 1980s, the low computing power offered by processors called for the creation of dedicated processing architectures initially based on multiprocessor architectures, composed of standard microprocessors (Dickmans BVV1 and BVV2 architectures, 8085 and 8X86 processors).

However, these architectures very quickly revealed their limits, notably in the performance at the level of inter-processor communications.

So by the end of the 1980s, research teams had already shifted to a new family of processors based on communicating processor architectures, namely transputers. These processors natively have a 20Mb/s data exchange connection. The Prolabs demonstrators had to have on-board machines of this type.

So the laboratory defined its own processor architecture, Transvision [10], a machine dedicated to image processing. This machine made up of 24 transputers enabled image capture directly in the memory of the processors.

So these parallel machines necessitated the study of algorithm parallelization methods, with a view to establishing on-board applications that met severe time constraints. With regard to the laboratory, the emphasis was on vision algorithms and pixel-related processing, which use up a lot of computing power. The theoretical assessments of speed (ratio between sequential time and parallel time for P processors) and efficiency (ratio between acceleration and the number P of processors) were checked experimentally, notably on algorithms for segmenting images in the sense of contours or region.

Two types of parallelism were then studied:

- parallelism by sharing the image, consisting in allocating the data to process between the processors. The processors execute the same programs on different areas of the image. This method speeds up the processing directly depending on the number of processors available in the event that the data is easily separable into independent blocks from the point of view of the algorithm to run.
- parallelism by sharing the algorithm to be set up, where each processor will execute a different task on different data. This type of parallelism is only possible if the processor is powerful enough to be able to process the entire task assigned to it in sufficient time. The interest of this type of parallelism is that it greatly reduces the amount of data to be exchanged between the processors.

3. Conclusion

In the electronics laboratory, the Prometheus project had a formative role by focusing the laboratory's work on activities tied to the automotive industry. During these years, the laboratory acquired an expertise on road scene perception systems, autonomous vehicle control or even the use of algorithms dedicated to artificial vision. The laboratory implemented its work on realistic demonstrators, notably on the Prolab 1 and Prolab 2 vehicles [19] (Fig. 4). French car manufacturers placed these vehicles at the disposal of the laboratories involved in the Prometheus project (ProArt group).

Figure 4: Prolab 2 instrumented demonstrator.

Like the story of the laboratory, the Prometheus project had a formative role in Europe, eliciting clusters in each country. Gallice/Alizon in France, Dickmanns in Germany, and Broggi in Italy were able to implement the first realistic demonstrators of driverless vehicles with the aid of national car manufacturers (Fig. 5).

Figure 5: Deutsch, Italian, French and English on Fiat test tracks (Turin, 1991).

Part II

1. Introduction

The pioneers were able to show that it was becoming conceivable to guide vehicles automatically, essentially like a driver does, based on vision data. In the second half of the 1990s, technological developments made it possible to refine further the concepts and to bring new solutions for steering and obstacle detection.

This period was marked by the advent of digital technologies that increased significantly the perception capacities of vehicles. For instance, the release of GPS differentials with centimetric precision, inertial units or even digital cameras added to the momentum created by the Prometheus project. Following this momentum, the laboratory expanded and became LASMEA, a combined research unit under the authority of Blaise Pascal University and the CNRS.

Firstly, the significant technological advances that enabled improved performance of on-board algorithms will be presented. Secondly, a new generation of on-board algorithms in vehicles based on these new technologies will also be presented.

2. Towards Technological Maturity

2.1 DGPS System

The American Navstar GPS system or GPS in short is currently the most developed satellite positioning system. It was developed by the US Department of Defense from the 1970s onwards. The first satellites were placed in orbit as of 1978 and GPS was declared fully operational on 27 April 1995.

Basically designed for military applications, the precision for civil applications was limited until 2000. The precision of a natural GPS system then went from roughly a hundred metres to about ten metres.

This precision is quite insufficient for driving a vehicle. Therefore systems, based on stations on the ground with a known positioning, helped to improve this precision. For instance, the American WAAS or European EGNOS systems use a series of ground stations to process parameters sent to mobile devices via geostationary satellites. As the position of the ground stations is known, a processing centre can estimate the corrections to be applied to the measurements and transmit them to the receiver via geostationary satellites.

The GPS RTK (Real Time Kinematics) system, by measuring the phase of the signal supplied by the satellites of the GPS constellation, and with the aid of a base station, is capable of providing a position in real time known with a standard deviation of about 2 cm. So this system consists of a base with a radio transmitter, which sends the differential corrections to a mobile receiver capable of sending its position measurement 10 times per second to the vehicle. This solution requires the reception of at least five common satellites between the base and the mobile device. This can be problematic notably in urban areas. That is why the first applications concerned the farming world for the most part, although many trials were run for driving other types of vehicles.

2.2 Radar System

The first experiments in the field of automotive radar already took place in the late 1950s. In the 1970s, more or less intensive radar developments started at microwave frequencies.

Accompanied by the remarkable progress in semiconductor microwave sources and in available computing power of microcontrollers and digital signal processing units, the commercialization of automotive radar became feasible in the 1990s. The radar systems for road transport have important advantages. Indeed, unlike the data from laser range-finders, radars are barely affected by weather conditions and through the Doppler effect they provide a direct and precise estimate of the speed of the targets.

2.3 Cameras

Until 1995, the use of artificial vision systems was hampered by the technologies required. As the images from the camera were transmitted in analogue (mostly in PAL format), the computer system had to do the analogue-digital conversion of the images from this video stream. This type of system meant that the quality of the digital images obtained was mediocre, with little contrast and a lot of noise. The geometric stability of these images depended directly on the quality of the PAL synchronization signal and could vary at times depending on the temperature of the components of the data acquisition board. Furthermore, the image resolutions and framerate were set by the PAL system (576 lines and 25 fps (frames per second)).

In 1995, the FireWire IEEE1394 bus (i-Link at Sony) revolutionized image acquisition channels. In fact, the transmission of images between the cameras and the computer system became digital, doing away with the analogue/digital conversion stage on input to the computer system. It was then possible to connect a camera to a standard PC at a lower cost. The direct consequence was the improvement of the geometric stability of the images. The data transmission protocol also guaranteed a very good temporal stability of the channel between the camera and the system.

3. Towards Scientific Maturity

3.1 Road Scene Analysis Systems

The group's efforts during these years focused on two essential functions: the detection and localization of other vehicles and the lateral localization of the vehicle on its traffic lane. At the time the aim of these studies was to be able to warn drivers in the event of dangerous situations and also to create systems like Adaptive Cruise Control (ACC) that adjust speed in relation to the situation ahead.

For the lateral localization of the vehicle in relation to traffic lanes, the initial vision solution was developed and a general approach allowing operation in varied contexts was developed. The road recognition was done by a detection technique guided by a road model. This model could be extended to roads with one or more lanes, or even without ground markings (Fig. 6).

Figure 6: Vision based road following.

Following on from the work done with the Lidar 3D detector, LASMEA had a 77GHz radar transceiver to run tests on the detection and tracking of obstacles. A major effort was made on the tracking methods.

Jouannin [11] suggested merging the radar, vision and proprioceptive detector (speed, turning angles) data by extensive Kalman filtering to obtain a more secure, solid positioning of the obstacles.

These methods were applied to measurements coming from a radar or from a camera observing vehicles equipped with visual markers [12] (Fig. 7). By linking this technique with the lateral localization, it was then possible to obtain an ACC-type function. In fact, the obstacle vehicles can be replaced on their traffic lane and finally determine which is the most dangerous. These studies were able to be tested in real time on board the VELAC demonstrator vehicle, which the laboratory acquired in 1998.

3.2 Systems for Precise Vehicle Control

The purpose of precise vehicle control and control law design is to calculate vehicles steering angles so that it follows a given trajectory. Knowing the location of a vehicle and the trajectory to follow, the system had to calculate the turning angle of the wheels to stay on course comfortably

Figure 7: Road following and obstacles detection.

and precisely. The required steering precision was in the region of a centimetre. Until then the techniques developed for steering vehicles were directly tied to the on-board perception capacities and notably the image processing algorithms. With the advent of RTK-DGPS, the detector could give the vehicle's positioning directly. As early as 1997, LASMEA acquired a GPS RTK system. The precise position of the vehicle was then directly provided by the detector and the vehicle's trajectory could be estimated through temporal filtering. This information made it possible to validate a vehicle control based on the exact kinematics model. The control laws were tried and tested at first on a combine harvester and a tractor. These tests established the feasibility of lateral guidance of machines by means of a single RTK-DGPS antenna on relatively flat farmland [13] (Fig. 8).

Figure 8: DGPS guidance of a farm tractor.

3.3 Systems of Localization by Vision

During this period, the vision localization technologies on board the vehicles relied on simple primitives such as following white lines, mowed/unmowed areas. However, by virtue of the improved quality of the cameras, a significant part of the group's scientific activities was devoted to developing metrology tools by vision. For instance, the emphasis was placed on camera calibration and automatic calibration, localization of rigid or articulated objects and reconstructing volumetric objects. The purpose of camera calibration was to turn a camera into a metrology tool.

With regard to camera calibration, radial distortions were factored into the models, so that the metrology could be done even with cameras with a short focal length (Fig. 9).

Figure 9: Before/After distortions estimation.

The calibration process was also simpler to implement because of the use of less precise calibration patterns [14]. In fact, the 3D coordinates of the landmarks used in the calibration patterns are included in the parameters to be estimated.

Moreover the search for discriminating landmarks was used to do the camera/object positioning. The methods used were based on matching points of interest located on the surface of the objects. To do so, icons were used to make correlations and then to recognize the texture of the area, to link the points in the image sequences. By filtering these points of interest and through triangulation, it was then possible to estimate the position of the objects.

The 3D structure of the space in which travels a camera can be calculated from this work. It is also possible to calculate the movements of a camera in that space. The bases for developing systems like Structure From Motion to calculate the movement of vehicles by vision were mastered.

For instance, if a camera was on board a vehicle, this system served to calculate the trajectory followed by the vehicle and the 3D structure of the environment surrounding the vehicle.

4. Conclusion

The technologies conceived by the pioneers were starting to reach maturity. Indeed, on-board road radars were marketed as of 2000, to provide an ACC function.

In the laboratory, the performance of control algorithms was mastered. The bases for developing systems like Structure From Motion to estimate the movement of vehicles by vision were known.

The problem of urban travel requires, among other things, finding an alternative to DGPS for localization. In fact, under the effect of urban canyons, the DGPS localization solution does not work well in town. So an effort went into finding an alternative vision solution for the localization of vehicles.

In the early 2000s, LASMEA decided to concentrate its work on the urban travel issue in the area of individual public driverless vehicles, while continuing its activity in the road transport field (notably through the ARCOS, PAROTO project). In 2002, the laboratory acquired a first test vehicle, the Cycab (Fig. 10).

Figure 10: The Cycab.

Part III

1. Introduction

The possibility of automating transport networks offers perspectives for reducing the harmful effects caused by urban transport systems. But public transport cannot satisfy all the users' travel needs. The creation of individual public transport would complement public transport to reach areas not covered or for late hour travel. These new modes of public transport could reduce harmful effects compared with the private car thanks to vehicles specifically adapted to an urban context: smaller, lighter, lower consumption of energy and parking spaces. At present, the only complement in terms of public transport is the taxi, but it is expensive.

Research projects like Bodega 2 and MobiVip 3 were launched to study the feasibility of non-instrumented automated individual public transport in urban areas. In this context, the challenge to overcome for the automation of vehicles is far greater than for the metro, because vehicles have to be able to travel over an extensive area, not just over a secure line equipped with sensors. So the objective was to produce a system of automatic shuttles, operating with no modification of the existing infrastructure, able to function like a horizontal lift, for example.

In the laboratory, in the early 2000s, the emphasis was on producing a localization system for a robot to navigate autonomously using a single camera, without adapting the urban infrastructure [15]. After some scientific and technical developments, realistic demonstrations showed the quality of the localization and control system, as well as its limits.

In the second half of the last decade, the complete chain from the camera up to the vehicle was optimized. During this period, a second camera was used, contacts were made with Ligier Group to create a vehicle really suited to the target service that could also be mass-produced. This work eventually led to the experimentation of a complete on-board system in the EZ10 vehicle.

2. Development of the Guidance System by Monocular Vision

The vision guidance systems developed until then consisted in following specific points of the environment like lines, edges of pavements or even lighting cues. So the idea developed consisted in creating in advance its own map or visual memory containing the specific elements of the environment that can serve as cues.

The principle retained consists in building a three-dimensional map of the area covered during a learning run, then using the map to locate itself (Fig. 11).

The "mapping" stage is an algorithm for 3D reconstruction from a moving camera (structure from motion). The map is a 3D model of the scene observed composed of a scatter plot. In this way, by doing a first run driving the vehicle manually, a map containing these elements is created, the trajectory followed by the vehicle is computed.

Knowing this map, an image taken with the same equipment in the same environment serves to locate the robot in relation to the environment.

The localization of the robot from an image consists of resetting the current image on the 3D model of the scene, from which we can deduce the position of the camera and therefore the robot (Fig. 12).

This localization can then be used to control the robot and ensure that it autonomously follows the same route as during the learning run.

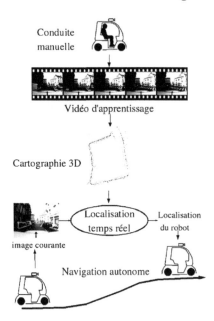

Figure 11: Learning and location process.

Figure 12: Location in city.

2.1 Map Building and Localization

So the first stage of the algorithm consists in detecting particular elements (primitives) of the environment, which can be found during successive runs of the vehicle. For this algorithm, a large number of points of interest (landmarks) (about 1000 detected in each image) is used for greater robustness. Several days may elapse between learning and localization, which means that certain elements present on the scene will be modified (Fig. 13).

Figure 13: Builing the map with snow/location without snow.

A part of the memorized points will no longer be usable so more points will need to be detected than the necessary minimum to locate itself. This leads to using a detector of quickly-computed points of interest based on the detection of corners in the images [21].

Combined with a standardized centred quickly-computed correlation calculation even if there are many points, it became possible to recognize these points all along the robot's route, then through triangulation to calculate their position in space, and to calculate the movements of the camera.

For the localization stage, this correlation calculation makes it possible to match the points from the 3D map with those seen by the camera. Once these matches are made, the position of the camera in the environment can be estimated.

2.2 Control

During these years, the control laws developed in natural surroundings were extended to the urban environment. The tests done with the GPS-RTK show that it is possible to follow a trajectory with great precision.

Once localization by vision became functional, the link between the two algorithms was made [16]. The performance on tracking trajectory was soon very good, thereby validating the precision of the localization by vision. Compared with DGPS, vision provided a much better estimate of course error. The precision of lateral error was virtually at the same level as DGPS. The vision algorithm also showed a satisfactory reliability with regards to variations in lighting conditions or instability in the environment. For example, it was possible to run tests with and without snow on the ground.

2.3 Real-World Test

2008 was a pivotal year for the project. In fact, during that year, the system was able to be tested under real-world conditions for passenger transport such as at the Airbus site in Toulouse. The difficulty lay in transporting people working on the site, notably at peak times in the morning, at midday and in the evening. The trajectory was about 1.5 km long. From a vision standpoint, it was necessary to split the trajectory into three parts, because the guidance computer memory could not hold the data required to drive over such long segments. This imposed stops of a few seconds to reload a new mapping at two points on the trajectory. Furthermore, as the trajectory ran east to west, there were difficulties with low-angle sunlight in the morning and evening. The demonstration was nonetheless a success since the three vehicles in operation travelled nearly 1,000 km automatically. The approach, on large private sites, clearly showed its usefulness. It could notably improve safety by drastically reducing the number of private cars at the site.

In 2008, as demands for experimentation in realistic conditions became increasingly urgent, the laboratory acquired a space called PAVIN (Plateforme d'Auvergne pour les Véhicules Intelligents – Auvergne Test Track for Intelligent Vehicles) to test driverless urban vehicles (Fig. 14).

Pavin is a 5000 sq m site located at LASMEA dedicated to urban vehicle experiments. It offers small vehicles the possibility of developing in realistic "urban" environments on tarmacked roads lined by buildings with crossroads, roundabout, traffic lights or even a passenger loading zone.

At the same time, contacts were made with Ligier. The objective of the FUI VIPA project was to create a vehicle for automatic passenger transport that could be mass-produced. A second objective was also to optimize the steering and improve reliability in case of low-angle sunlight by adding a second camera to the vehicle.

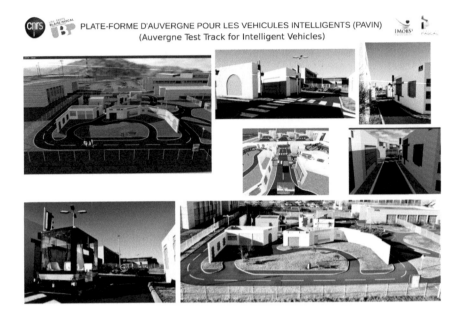

PLATE-FORME D'AUVERGNE POUR LES VEHICULES INTELLIGENTS (PAVIN)
(Auvergne Test Track for Intelligent Vehicles)

Figure 14: Virtual and real PAVIN – Vipalab vehicle.

3. Development of the Guidance System by Bi-Camera Vision

From 2008-2011, LASMEA put its efforts into the VIPA (Véhicule Individuel Public Automatique) project in partnership with the innovative SME Apojee and Automobiles Ligier [lig] to develop a vehicle providing a function like the horizontal lift. The VIPA vehicle was designed to be produced in medium run, marketed and used in urban areas, such as car parks, pedestrian precincts, serving a hospital or airport terminal from a distant car park, automatic tour of tourist attractions or amusement parks.

The use of a single camera was not robust enough under outdoor lighting conditions; notably due to overexposure problems (occasionally encountered at sunrise or sunset or else on leaving a building to enter a sunny area). It was to solve this problem that the VIPA vehicles were equipped with two synchronized cameras, one in front (facing forward), and the other in the rear (facing backward). In this way, if one of the cameras was dazzled by the sun, another camera could provide the relevant data. Besides the symmetry of the vehicle, another advantage of this multiple camera system was that it provided redundant information and therefore made the localization more robust [17, 18].

3.1 Bi-Camera System

The learning and localization techniques were expanded to a bi-camera system (Fig. 15).

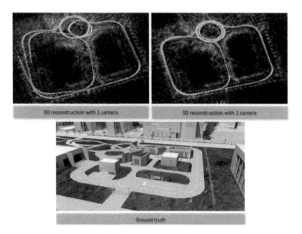

Figure 15: Reconstruction with one or two cameras.

But in order to be able to rebuild the 3D map, use the localization algorithms, the multiple camera system first needed to be calibrated. The calibration then consisted in estimating the extrinsic settings of the cameras, the parameters for the rigid conversion between two cameras (i.e. their relative positions).

The work done at the time showed that extrinsic calibration could be obtained concurrently with 3D reconstruction (for example during the learning stage), by using points of interest as landmarks [20].

During the calibration procedure, the vehicle must complete a trajectory including a return trip. During the return trip, the landmarks seen earlier by the front camera are observed by the rear camera, which serves to produce front/rear pairings. The scene and the positions of cameras are then optimized, thus enabling an estimate of the extrinsic parameters of the cameras too. The method developed is easy to use because it does not necessitate a mirror, or any *a priori* about the speed of the multiple camera system, or full knowledge of the geometry of a target beforehand.

3.2 The VIPA Vehicle

A new clean autonomous mobility concept, VIPA (Véhicule Individuel Public Autonome) is an electric driverless vehicle that can carry up to six passengers – four seated and two standing – over short distances, at a top speed of 20 kph at the most (Fig. 16).

Figure 16: The VIPA vehicle.

The vehicle was first presented in a static way at the 2010 Paris car show, then in motion during the Michelin Challenge Bibendum 2011 in Berlin.

From September 2013 to February 2014, the VIPA was tested by users and staff of the Estaing teaching hospital in Clermont Ferrand. This was the first long-term, full-scale experiment, enabling a running-in period needed to optimize its functions. During those six months, the ergonomists and psychologist of the LAPSCO laboratory assessed reactions of users. The challenge is to quickly move from a pilot vehicle to an industrializable solution.

This demonstration showed the difficulty in integrating the system on a site. In fact, the vehicle drove on the pavement in a very confined area. That imposed a very precise calibration of the steering mechanisms to be able to meet the specifications, and a slow driving speed for the vehicle because the edges of the operating zone were detected by the vehicle's pedestrian safety system. Furthermore, the vehicle used the space previously used by pedestrians, which made cohabitation difficult.

In the light of these facts, automatic driving has shown that it can be superior to a driver by virtue of its repetitiveness: in fact, in manual driving, because of the narrowness of the path, it is virtually impossible to complete the entire trajectory without manoeuvre.

The acceptability tests were quite positive: the public were interested by the innovation aspect and participating in a full-scale test. The service aspect needs to be improved in a context like the teaching hospital. So the Ligier Group decided to design a new, more suitable vehicle: the EZ10 (Fig. 17).

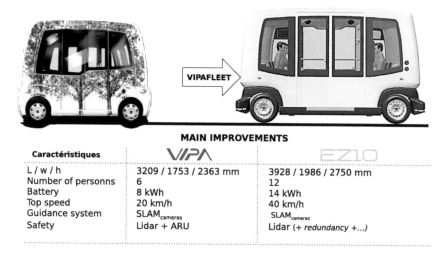

MAIN IMPROVEMENTS

Caractéristiques	VIPA	EZ10
L / w / h	3209 / 1753 / 2363 mm	3928 / 1986 / 2750 mm
Number of personns	6	12
Battery	8 kWh	14 kWh
Top speed	20 km/h	40 km/h
Guidance system	SLAM$_{cameras}$	SLAM$_{cameras}$
Safety	Lidar + ARU	Lidar (+ *redundancy* +...)

Figure 17: EZ10 vs VIPA.

This vehicle corrected the weaknesses of the 1st vehicle with regards to on-board comfort and accessibility. It can indeed be accessed by mobility impaired persons. It also includes two drive trains to improve its agility and increase its effectiveness for driving in a confined environment.

4. Conclusions

The EZ10 vehicle was first presented in a static way at the 2014 Paris car show, then in motion during the Michelin Challenge Bibendum 2014 in Chengdu. From June to September 2015, the EZ10 was tested by users and staff at the Michelin research centre in Clermont Ferrand. It was also the test vehicle for the European CityMobil project in the towns of Lausanne and Vantaa with a steering technique based on SLAM Lidar.

Final Conclusions

After the first EZ10 experimental tests, a new FUI Vipa Fleet project has started up. The VIPA Fleet project built on the continuity of the VIPA/ EZ10 projects, which produced a first version of the vehicle capable of evolving on simple circuits in "horizontal lift" mode (Fig. 18).

Vipa Fleet aims to continue improving the reliability and performance of the guidance system and the vehicle. The project also aims to make the vehicle communicating, in order to improve the service provided to users and to enable fleet management of automatic vehicles.

Figure 18: EZ10 vehicle.

So the aim of this project was to develop a 2nd optimized version of the transport system that can evolve notably:

- In a fleet of vehicles: interactions with other driverless vehicles on the site, with a fleet management server and call terminals. A fleet of five EZ10 should be able to function in Tramway or Taxi mode according to the demand.
- On more complex circuits in interactions with other vehicles or pedestrians present on the route.

This project served to further improve the autonomous vehicle concept and its reliability, to develop the system of communication and fleet management and to streamline the coordination of vehicles in a complex environment.

Thanks to the real-world experiment on a closed demonstration site, this project grasped the difficulties relating to:

- putting together a fleet of driverless vehicles and providing the necessary technical corrections before its industrialization
- the uses noted *in situ*: acceptability of the vehicle, design of interfaces, human, technical and organizational characteristics of the environment hosting the vehicles.

Vipa Fleet permits to test the EZ10 vehicle on a looped route of 1.4 km, comprising eight stations. The vehicle became communicating. So it can respond to driving instructions sent by the fleet management system. It is also capable of providing information to the fleet management system, like the level of its batteries or the kilometres travelled. The localization system was further optimized, making it possible to improve the guidance reliability and cover nearly 1500 km driverless, steering by cameras.

At the same time, the industrial structuring of the project continued. The guidance was developed and industrialized by Robosoft Driverless Solution (RDS). EasyMile was in charge of managing the servicing of these vehicles.

At the laboratory level, the research work continued in three key areas:

- localization and guidance in order to create a vehicle capable of locating itself with greater reliability, of developing in taxi mode and avoiding obstacles effectively,
- fleet management to efficiently manage a fleet of vehicles capable of responding to requests in real time,
- the acceptability of the vehicle in order to optimize the vehicle or the system and to be able to integrate to best advantage in its environment.

So the ultimate aim of this work is no longer to develop a driverless vehicle, but a complex communicating system so as to provide a transport service on a site, in close interaction with the users and the site's managers.

Acknowledgements

Photos in 1st and 2nd part courtesy of Joseph Alizon.
Figures 16 and 17 courtesy of Ligier Group.

References

1. H. Moravec, 1983. The Stanford cart and the CMU rover. *Proceedings of the IEEE*, 71, pp. 872–884.
2. S. Tsugawa, T. Yatabe, T. Hirose and S. Matsumoto 1979. An automobile with artificial intelligence. 6th IJCAI, Tokyo. pp. 893–895.
3. Dickmanns, E.D. and Zapp, A. 1985. Guiding land vehicles along roadways by computer vision. *Congres Automatique*, pp. 233–244. Toulouse: AFCET.
4. Electronic Lab. 1988. Activity Report (1985–1988). Clermont Ferrand.
5. Electronic Lab. 1992. Activity Report (1989–1992). Clermont Ferrand.
6. B. De Mathelin, J.A. 1992. Brevet n° EP Patent App. EP19,920,400,971.
7. R. Chapuis, J.G. 1991. Real time road mark following. *Signal Processsing*, 24, 331–343.
8. R. Aufrere, R.C. 2001. A model driven approach for real time road recognition. *International Journal of Machine Vision and Applications*, 13, 93–107.
9. F. Jurie, P.R. 1994. High-speed vehicle guidance based on vision. *Control Engineering Practice*, 2(2), 289–297.
10. J.P. Derutin, B.B. 1991. A parallel vision machine: Transvision. Computer Architecture for Machine Perception (CAMP), 241–251.

11. S. Jouannin, 1999, Association et fusion de données : application au suivi et à la localisation d'obstacles (data association and fusion : application to obstacles tracking and location), PhD Thesis, Blaise Pascal University, Clermont Ferrand. PhD thesis. Universite Blaise Pascal, Clermont Ferrand.

12. R. Chapuis, F.M. (2000). Road detection and vehicles tracking by vision for an on-board acc system in the velac vehicle. *FUSION 2000*, 2.

13. B. Thuilot, C.C. 2002. Automatic guidance of a farm tractor relying on a single cp-dgps. *Autonomous Robots*, 13(1), pp. 53–71.

14. J.M. Lavest, M.V. 1998. Do we really need an accurate calibration pattern to achieve a reliable camera calibration? European Conference on Computer Vision (ECCV'98), pp. 158–174.

15. E. Royer, 2006. Cartographie 3D et localisation par vision monoculaire pour la navigation autonome (Monocular vision for 3D mapping and location for autonomous navigation), Blaise Pascal University, Clermont Ferrand.

16. E. Royer, J.B. 2005. Outdoor autonomous navigation using monocular vision. IEEE/RSJ International Conference on Intelligent Robots and Systems (IROS 2005), pp. 1253–1258.

17. Pless, R. 2003. Using many cameras as one. *CVPR*, pp. 587–593.

18. Chang, M. and Y. Wong, K.Y. 2007. A robust head pose tracking system based on multiple cameras (Vol. Internal report number:khw_irep_070509). The Chinese University of Hong Kong.

19. L. Trassoudaine, S.J. 1996. Tracking-systems for intelligent road vehicles. *International Journal of Systems Science*, 27(8), 731–743.

20. Lebraly, P. 2006. Étalonnage de caméras à champs disjoints et reconstruction 3D – Application à un robot mobile. PhD thesis. Clermont Ferrand.

21. C. Harris and M Stephens. 1988. A combined corner and edge detector. In: Alvey Vision Conference, 147-151.

Tolling Systems: Towards a Global Set of Standards

Fausto Caneschi*

C/o Lecit Consulting, via Pasquale Landi 9
56124 Pisa, Italy

Abstract

A concise panorama of the history and the technologies for Electronic Fee Collection (EFC) is given, together with some criteria commonly used for evaluating the different technologies as used for EFC, and some examples of deployments. The importance and the role of EFC standards are presented, together with their current status, with a special attention to architectural aspects. Finally, the EFC is presented in the framework of Intelligent Transportation Systems.

WHICH ARCHITECTURES, TECHNOLOGIES, AND STANDARDS FOR EFC?

1. Architecture and Functions

A general architecture for Electronic Fee Collection systems is defined in the standard ISO 17573 (ISO 2010 in [1]), where the main roles in a tolling system are identified, as in Fig. 1.

The four roles identified in Fig. 1 represent the main actors in a tolling system: the user of the road, the toll charger, which is the entity entitled to apply a toll, the toll service provider, which interfaces the user by means of a contract and provides him with the technical means for electronic tolling and the manager of the tolling environment, which is typically (even if not always) a road authority.

---

*Corresponding author: fausto.caneschi@lecitconsulting.it

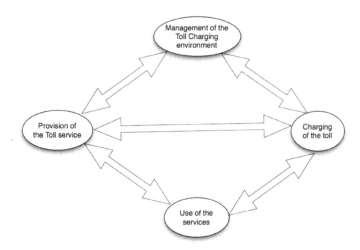

Figure 1: Roles in a tolling system.

Arrows in the figure show generalized interactions among roles: each role being in principle a separate legal entity, standards are in principle needed for each of those interactions. These interactions are achieved by means of functions.

A paper dealing with Open Road Tolling (ORT) [2] identifies main classes of functions within an ORT system, which, by the way, are valid for all EFC systems.

- Identification of the payer being tolled
- Identification of the vehicle being tolled
- Identification of the trip being tolled
- Definition of the transport rate (tariff) table
- Calculation of the amount due for a toll transaction
- Invoicing/payment/customer management
- Dealing with occasional users
- Control/checking (enforcement) systems

Each class of functions belong to different parts of the overall tolling system, as summarized in the diagram of Fig. 2.

Some functions clearly belong to one role, while others may be performed by different roles, either exclusively or in collaboration with other roles. The actor (or actors) that performs a given function and the way the function is performed, depend on the type of the tolling system. For example, in some systems the identification of the trip and even the calculation of the amount due (the toll) are left to the toll service provider, the toll charger being only involved in control and check (enforcement). In other systems, it is the toll charger that does all the job and the service provider is in charge of dealing just with its users and the invoices coming

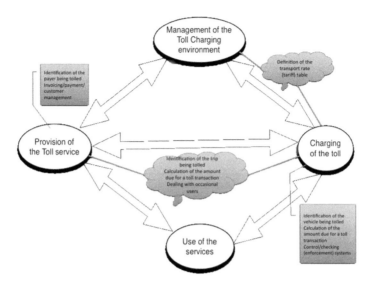

Figure 2: Roles and functions in a tolling system.

from the toll charger. A detailed analysis of these aspects will be found later on. Here, it is important to point out that the actual transfer of money from the user to the service provider, and, ultimately, to the toll charger, is completely out of the scope of the standardization activity on electronic toll collection.

2. Functions and Standards

In the real world, not all the identified (classes of) functions need to be standardized. First of all, the entity that manages the overall tolling system may vary from case to case, and sometimes it is not even there as a juridical person. As an example, very often the toll charger (entity that "charges the toll") is a concessionaire who operates on a contract on behalf of a state owned Road Authority. Relations, and, consequently, interactions, between a concessionaire and a Road Authority depend strongly on local laws, and cannot be subject to international standardization. As another example, this managing role could in principle be taken over by a super national entity, like for example the European Commission in the European Electronic Tolling Service (for information on the EETS, a good starting point is European Commission:2001 in [2]), but this is not at the moment implemented, and will happen, if ever, only in the future, while the EETS will supposedly already be operational.

Interactions between Users and Service Providers or Toll Chargers are also subject to laws (think of privacy laws, for example), which are different from country to country, and, again, difficult, if not impossible, to standardize.

What indeed is an area for standardization, is the whole set of interactions between toll chargers and toll service providers. Considering the diagram in Fig. 2, one will thus concentrate on the standardization of functions related to:

- Identification of the vehicle being tolled
- Identification of the trip being tolled
- Control/checking (enforcement) systems

Before looking at the standards, however, the technologies used to implement the above functions should be considered.

3. Technologies

Each class of functions eligible for standardization can be implemented by using a number of different technologies, which were devised and deployed in the last 40 years.

3.1 Vehicle Identification

Vehicles can be identified in multiple ways and using different technologies. Sometimes, vehicle identification is related to vehicle classification, even though the two aspects are different. One "natural" way to identify vehicles automatically is in principle Automatic Vehicles Identification (AVI), where basically two families of technologies are identified, namely Dedicated Short Range Communication (DSRC, based upon an application layer defined in EN 12834 in [6]), and Radio Frequency Identification (RFID, defined in the series of standards commonly referred to as ISO/IEC 18000 in [4]). These two technologies are functionally similar, the main differences being the operating frequency range and the fact that the former is able to deal with dynamically computed data, while the latter uses basically static data. Both are based on passive or semi-passive devices, which receive most, if not all, the energy they need[*] from the antenna via radio transmission and are able to use that energy to respond to an interrogator by providing data stored in a chip. However, identification via AVI is not much used in real life tolling systems. DSRC is often used to identify vehicles by means of tolling transactions and by

---

[*] The battery that is used in most DSRC devices is to ensure some additional processing capabilities related to tolling, but the basic functioning principles are the same. This is why these devices are often called semi-passive.

associating the vehicle itself to the on-board device used for tolling*. Also, Automatic Licence Plate Recognition (ANPR) is often used to complement DSRC or RFID identification.

Finally, technologies used to classify vehicles, such as lasers, to measure inter-axle distance or height, or inductive loops buried in the road, might also be used to identify vehicles, even though they are used mostly to classify them in order to determine their tolling classes. These latter technologies are also subject of international standardization, but not strictly in the scope of tolling.

3.2 Trip Identification

If today one thinks of how to identify the trip a vehicle has done on any route, there is an immediate answer, that is "use a GPS". However, GPS alone is not enough for tolling. You need to communicate further the positioning information to an entity that computes the toll or, if you want to be able to compute the amount of the toll "on board", to be able to receive the toll domain characteristics (including tariffs). This implies either using a communication media (that is normally the cellular network), or an offline download of the positioning data to a server station. All this, of course, works in a perfect world where nobody cheats (by tampering with the positioning device), and where the GPS signal is always present, including tunnels and urban canyons. The former aspect is covered in the Enforcement section. About the latter, additional information needs to be available on the road to be downloaded into vehicles by some kind of communication.

In actuality, one does not need a precise and continuous monitoring of the position of a vehicle to calculate a toll, as a GPS would do, at least not in all kinds of tolling systems and toll schemata. When the tolling system consists of a set of roads where barriers regulate entry and exit and possibly passage, one has a set of fixed tolling points where it is the real toll transaction that indicates the position of the vehicle. That is the case of DSRC tolling.

In conclusion, positioning technologies are based on either:

1. The principle that infrastructure should be kept to a minimum
   Here, GPS (or inertial navigation) is used. These technologies, to be used for tolling, must be complemented by other means such as a communication link with control centres to ensure proper update of characteristics of a tolling domain on on-board devices and proper transmission of positioning data from on-board

---

* This implies that each vehicle has a unique tolling device, which cannot be the case in some tolling schemata, where a tolling device can be associated with more vehicles.

devices. In addition to that, the so-called "location augmentation" capabilities must be present on the territory, to calibrate the positional data in vehicles, as well as cover situation where the GPS signal is not present.

2. The principle that on-board devices should be simple and not expensive

   Infrastructure installations can be kept at a reasonable level of complication (and cost) due to the characteristics of the tolling domain. In this case, positioning of vehicles is determined by the execution of tolling transactions at pre-defined points, whose location is known.

3.3 Enforcement

As stated before, in an ideal world, enforcement should not be needed. In actuality, people cheat and toll chargers do not like to lose money. So, enforcement techniques are needed, whose sophistication and complexity vary with the characteristics of the tolling system.

When many critical functions are brought inside the on-board device, such as positioning or even toll calculation, one needs means to verify that the device has not been tampered with. Thus, to avoid stopping vehicles and plug a test unit in the on-board device, one needs to ensure that the data recorded by the device is trusted, and means to verify that are needed. The former can be implemented by using so-called "trusted recorders", which are certified units installed in the on-board devices, that store positioning data (and other information) in a secure environment. These trusted recorders can be interrogated remotely to download data and, at the same time, to ensure that the on-board unit works correctly indeed. A simplification of this model avoids using trusted recorders, but just performs a remote interrogation which, by elaborating the data received, enables discovering a possible tampering of the on-board unit.

When the infrastructure is in charge of performing almost all tolling functions, enforcement can normally be done by using laser triggers and cameras that are able to spot vehicles that physically elude a tolling or a control point (avoiding barriers is more difficult, but it is done at a higher frequency that one can expect).

3.4 Technology Deployment

The technologies described above, when deployed, have an impact on the two main physical entities in a tolling system, i.e., vehicles and road infrastructure. Some (most, one could say) of these technologies have been standardized as per either local or global norms, and a nice challenge is to get through the jungle of standards to decide which technologies (and consequently which standards) suit one's needs best. To do that, criteria must be devised to classify those technologies. A classification criterion can be applied, that is the impact of the deployment of a technology (as seen as the implementation of a standard) on vehicles vs. its impact on the infrastructure, where by "impact" one can use any metrics such as cost, complexity of implementation, and so on. It can be stated qualitatively that the two impacts follow a hyperbolical law, i.e., the product of the two is a constant, representing the level of functionality of the tolling system (see Fig. 3). By functionality one simply means here a global indicator that combines all functions offered by the tolling system[*] (easiness of use, efficiency, performance, …). To ensure the same level of functionality one thus has to compromise between the complexity of on-board devices and the complexity of the infrastructure. This complexity of technology has a clear impact on the type of standards that can be used.

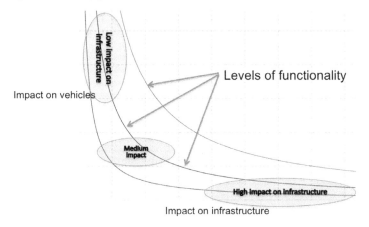

Figure 3: Technology impact on vehicles and infrastructure.

Given the hyperbolic law shown above, one can discuss the impact of standards implementation from either the vehicles' viewpoint, and from that derive the impact on the infrastructure, or vice versa, indifferently. From the point of view of vehicles, three main categories of impact can be identified.

1. **Low or zero impact:** These are all the technologies that rely on infrastructure devices able to identify and classify vehicles with a weak or even zero interaction with vehicles, such as video tolling systems (APNR), or a one-way information communication such as that with a passive RFID sticker.
2. **Medium impact:** These are all the technologies that imply a two-way interaction between an on-board device and the infrastructure. In this

---

[*] One is not going to discuss this concept further as it is not essential for this paper.

case, the on-board device has to have some intelligence and memory, but just limited to the exchange of data needed to, e.g., communicate its identity and a few more static or dynamic information (information computed at interrogation time). Examples of these technologies are those based on short distance radio communications, such as DSRC, operating in the GHz range of frequencies, or infrared.

3. **High impact:** These are all those technologies where the infrastructure is not strictly involved in the determination of the toll, so that in some cases it is said that there is no need of infrastructure*. As a compensation, the burden is all in the on-board device. Tolling systems using these technologies are often named autonomous tolling systems, meaning that the on-board unit can perform its job autonomously, i.e. with no interactions with the infrastructure. Examples of these technologies are GNSS-CN (Global Navigation Satellite System – Cellular Network), where positioning is achieved by means of GPS location and communication between vehicles and central systems is done via cellular network and tachograph, where positioning is achieved by means of either GPS, or inertial navigation systems, or a combination of both.

The following table gives a summary of the technologies that can be used reasonably today for EFC (because of their availability, cost, impact, historical usage, ...), according to their impact on vehicles.

Impact on vehicles	Vehicle identification/ classification	Trip identification	Control/ checking
No impact	• APNR • Passive RFID • Inductive loops	• APNR • Passive RFID • Inductive loops	• APNR • Passive RFID
Medium impact	• DSRC • Infrared communication • Inductive loops	• DSRC • Infrared communication • Inductive loops	• APNR
Full impact	• Cellular Network communication • Physical connection	• GPS positioning • Inertial navigation	• APNR • DSRC • Infrared

The categories shown in the table above are not completely independent of each other. Even if somebody could think in principle of using technologies that have different impacts (rows in the table) for each class of functions (columns in the table), in practice this is not economically sound, because the choice of a technology for one class of functions gives the possibility of using the same technology for other classes of functions. For example, once one has decided to use DSRC for vehicle identification and classification (and, consequently, toll

---

* This is not completely true, as has been shown before.

calculation), the same infrastructure and on-board devices can be used, according to the characteristics of the toll domain, for trip identification, so there is no convenience to add a different technology. In essence, the main choice one has to make is on the technology impact (choosing one row of the table). Once this is done, detailed technology choices are almost consequential. So, the question is rather on the model of the tolling system to implement (top down approach), rather than the technology one has to use (bottom up approach)*. However, technologies need to be evaluated for their suitability to given scenarios (that "almost" in the sentence above) and that is the subject of what follows.

4. Evaluation and Deployments

A paper of some years ago [8] made a comparison between different technologies for ETC and ITS, starting from the assertion that DSRC was at that time the preferred technology for ETC. At that time, infrared communication was seen as a promising "new" technology, while satellite-based tolling was seen as the technology of the future. A more recent paper [9] boasts the use of smart phones and Near Field Communication (NFC) to design a lightweight and scalable tolling system.

However, on the one hand, infrared technology had some initial problems with attenuation (or even cancellation) of the signal due to external factors (atmospherical conditions, for example), which slowed down its deployment, so the cost per unit became much higher than that of a DSRC based system. Eventually, when the technology reached a comparable reliability, the market, at least in Europe and in the US, was already set for DSRC. On the other hand, new technologies, like those based on smartphone systems, do not yet have the reliability and the security features of other systems. Finally, ANPR systems do not yet reach the more than 98% reliability[†] of, say a DSRC system. In the mean time, tolling systems have been deployed and running and at the moment, all new systems are based on the two paradigms that have been cited in this paper repeatedly: light on board units, which in practice means DSRC based systems, vs. light infrastructure, which in practice means satellite based (with a niche market for tachograph based) systems[‡]. These latter

---

* Unfortunately, most of the times the question one has to answer first when asked as a consultant is "which is the technology I have to buy?", even before one knows the whereabouts of the tolling system.

† In terms of successfully performed tolling transactions vs. the number of vehicles subject to toll.

‡ If the arguments put forward are not satisfying, one should think of Dante Inferno V: *"vuolsi così colà dove si puote/ciò che si vuole: e più non dimandare"*, as translated by Longfellow: "It is so willed there where is power to do/That which is willed; and ask no further question.

have been collectively named with the neutral term "autonoumous systems", where the adjective "autonomous" indicates that the On Board Equipment (OBE) can collect trip information (and more) autonomously, i.e., without interacting with an infrastructure on the road. The already cited paper of Malbrunot on ORT (Malbrunot 2006 [3]) gives some hints on criteria to use to decide for one type of system vs. the other one, based on number of vehicles and length of the roads. Although costs as expressed in that paper might need to be revised to updated figures, yet the methodology herein can still be used profitably. As standards follow the technology, one is thus going to see which are the standards currently available for electronic tolling.

THE STANDARDS FOR ELECTRONIC TOLLING

5. Computational Objects and Their Interfaces

As hinted at times in this paper, standards and technology can very often be seen as the same thing. If one thus wants to see which the relevant standards for electronic tolling are, one has to move from the diagram of Fig. 2, where roles and activities are depicted, to a more engineering point of view, meaning that one has to expand (or to view from another perspective) that arrow showing the interactions between toll charger and service provider. In other words, one must consider that those interactions should not be seen as limited to direct information transfers, where data are physically transferred between the two entities. Additionally, other indirect information transfers happen, e.g. whenever a tolling transaction is performed between an infrastructure device belonging to a toll charger and an on-board unit belonging to a service provider, or when data describing the geography or the fee structure of a toll domain, that is ultimately coming from the toll charger, is transferred to an intelligent on-board unit by the service provider. Besides, as the arrows in Fig. 2 are independent of the technologies that are used, the view should include all possible technologies. Summing up, as one is now dealing with real entities[*], one has to figure out which the computational objects are (for an accurate description of what a computational object is, one can refer to ISO/IEC:1998 in [10]) that model the toll charger-service provider interactions, and which their interfaces are.

Four isolatable computational objects can be identified, at the borders of which one can see standardizable interfaces:

1. The toll charger, represented here by its "Central System". By Central System it is meant here that the whole complex of

---

[*] Technologies apply to reality, it should go without saying.

computing facilities are under the responsibility of the toll charger and act in the scope of tolling.

2. The service provider, also represented here by its Central System
3. The Infrastructure, here representing the equipment deployed in the territory for tolling purposes, and collectively named Road Side Equipment (RSE).
4. The on-board equipment (OBE), here representing the equipment deployed in vehicles.

These four objects and their interactions are depicted in Fig. 4.

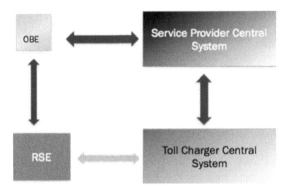

Figure 4: Computational objects in EFC systems.

In the diagram, three darker arrows indicate that standards can be (and have been) developed for the related interfaces. One lighter arrow indicates that no public standards are defined for those interfaces. The reason of this is simply that the interaction between RSEs and toll charger Central Systems is an internal affair, that happens between devices belonging to the same entity and therefore not worth standardizing.

This ideal model has been corrupted partially by a long standing discussion on the role and the functions that on-board equipments should play, also known as the "dispute on the fat OBE" that happened when devising the standards for autonomous systems. The more functions the OBE performs, the less the burden on the central system is, but that also has an impact on the engineering of the interfaces and related standards. Two extreme views were on the table: one to have a very simple OBE that simply collects positioning data and sends them to a central system, and the other where the OBE knows the characteristics of the tolling system, including the fees, makes all calculations and just communicates the amount to be paid. To solve the question, the concept of Proxy was introduced. If a logical device that is made of one component inside vehicles is imagined, and another component (the Proxy) that stays still somewhere and presents to the service provider an interface such as that a "fat" OBE would sport, then one can standardize that interface and leave the decision of where to put functions to implementors. This concept is represented by modifying the diagram in Fig. 4 to that one shown in Fig. 5.

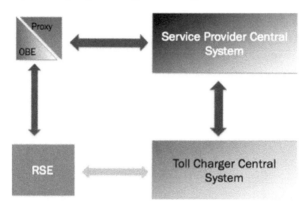

Figure 5: Actual computational objects and interfaces.

The OBE/Proxy object has an internal interface that represents the subdivision of functions between the two physical entities. That interface is not subject to standardization. Instead, two external interfaces are standardized, one with the service provider Central System, and another one with the toll charger's infrastructure, represented by the RSE object. Finally, the direct interface between the toll charger and the service provider Central System is subject to standardization.

All the standardized interfaces are defined, and for most of them the second edition of the related standards is ready. Figure 6 lists the names of the most diffused available standards for each interface.

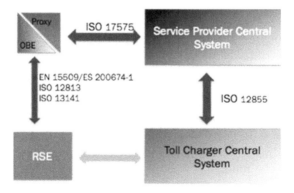

Figure 6: Standard interactions between toll charger and service provider.

The diagram in Fig. 6 represents an "impossible" system, i.e., one that uses both autonomous and DSRC OBEs to perform tolling. While this scenario is not forbidden, its implementation would certainly mean a waste of resources. The two tolling system models referred so far, namely, the infrastructure-based and the OBE based, are two cases of the general picture of Fig. 6, which will be dealt with in what follows, together with a short description of the cited standards.

5.1 Autonomous OBE Tolling Systems

In a tolling system based on autonomous OBEs, Fig. 6 will be simplified as in Fig. 7.

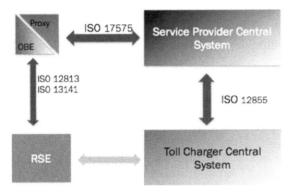

Figure 7: Autonoumous OBE tolling system.

On the one hand, the service provider Central System interacts with the OBE/Proxy to provide information about known tolling systems such as the structure of tariffs and the tolling domain layout (which segments, or areas, or sections are subject to tolling in terms of geographical coordinates). This information is provided by the toll charger at contract establishment time, and updated in due course. On the other hand, the Central System receives from the OBE/Proxy the amounts to be paid, together with the related evidences (traversed sections, times and all other related tolling information). This tolling information is collected by the OBE/Proxy when traversing a tolling domain, processed using the toll domain layout information and tariffs, and given to the Central System either on demand or at predefined events. The two interactions described above, in terms of data definitions, are defined by the ISO 17575 [11] standard. The standard, however, does not define the technical means by which this information is physically transferred to/from the Central System from/to the OBE/Proxy. The reason, again, stems from the possible varieties of the OBE/Proxy pair. If, for example, the Proxy disappears because the OBE is "fat", then a communication between that OBE and a Central System can better be done via OTA (Over The Air programming), like for cellular phones. If, conversely, the OBE is "slim", then Proxy is a real computing system and the major part of information transfer between the Proxy and the Central System can be done by using traditional file transfer mechanisms.

However, independent of how fat the OBE is, the tolling system cannot work without a little help from two infrastructure friends, which are represented in the arrow showing the interactions OBE-RSE. The first helper is to improve the positioning system of the OBE by transmitting accurate positioning information via a DSRC link. Depending on the geographical characteristics of the toll domain, the toll charger can deploy RSEs that interact with OBEs giving them the coordinates of their position. These coordinates (which are certified and digitally signed to avoid fakes) are then used by the OBE to measure the accuracy of the GPS, or to complement its information when the GPS signal is insufficient. The way positioning data is transferred from the RSE to the OBE via a DSRC link is specified in the Location Augmentation Communication (LAC) standard (ISO 13141, see [16]). Also, even if toll chargers may trust any service provider, yet they may want to control, on a statistical basis, or even systematically, that the OBEs that are traversing their territories are indeed worth of that trust. Controlling OBEs mounted on moving vehicles without interrupting the traffic flow can be done, again using a DSRC link and reading some parameters. This is specified in the Compliance Check Communication (CCC) standard (ISO 12813, see [15]). Both these standards use certification algorithms to prove to the OBE that the positioning transmitter in the case of LAC, or the interrogator in the case of CCC, are authorized devices. The way these algorithms work will be explained later on.

Once the relevant information for tolling has been collected and computed, payment settlements take place between the toll charger and the service provider. This information exchange, together with all direct information exchanges between toll charger and service provider, is defined by the ISO 12855 standard. As this, however, is common to all tolling system models, it will be dealt with in a separate clause.

5.2 Infrastructure Based Tolling Systems

Infrastructure based tolling system are adopting the DSRC communication technology more and more. There are two variants of this technology, one based on radio transmission around a 5.8 GHz carrier, the other one using a 5.9 GHz carrier*. This latter one is used for tolling only in North America, and with several variants. International standards are available

---

* Other frequency bands are, or have been used, but limited to single cases, and going to disappear.

for the former one, which is at the moment the technology of choice, so that from this point on only this will be referred to.

In a tolling system based on DSRC OBEs, Fig. 6 will be simplified as in Fig. 8.

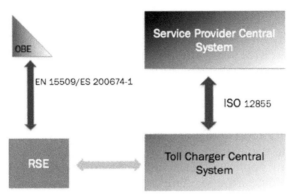

Figure 8: DSRC based tolling system.

The first notable difference between Fig. 7 and Fig. 8 is the vanishing of the Proxy and the related standardized interaction between it and the service provider Central System. This, of course, does not mean that standards prohibit interaction between service providers and OBEs; just those interactions, if any, are not standardized for an infrastructure based tolling system. One is in this case facing two standard interactions:

1. between RSEs and OBEs, providing for positioning, identification, and collection of all toll relevant information;
2. between toll chargers and service providers, providing for all other needed information exchanges. As anticipated earlier on, this interaction will be dealt with separately.

The DSRC tolling transaction that happens between OBEs and RSEs is regulated by a number of standards, the three most significant in terms of installed units being (the list is by alphabetical order):

1. ARIB (not drawn in Fig. 8), the collective name given to a set of standards, of which T75 (see [17]) and T110 (see [18]) are the most relevant for tolling. The ARIB standards are used in Japan.
2. CEN TC278, again a collective name given to the DSRC standards developed by CEN, and adopted in ISO, of which the most relevant for tolling is the EN 15509 (see [13]). CEN TC278 standards are used in Europe (except Italy, see next item) and in various other countries outside Europe.
3. ETSI EN 200 674-1 (see [14]), which is a complete DSRC protocol stack, including a profile for tolling, that is in use in Italy.

The reasons why two standards are in use in Europe is historical. The Italian standard, subsequently re-defined in the cited ETSI standard, was the first to be deployed in Europe and the one with the single biggest deployment (at the time this paper was written, more than eight millions of OBEs are deployed in Italy). When the European Union decided to support a pan-European tolling system, the already mentioned EETS, the decision was to specify an OBE capable of running DSRC transactions based on both CEN and ETSI standards and at the same time be able to operate in an autonomous tolling system, so as to accommodate all different systems running in Europe by reconciling the differences in the on-board units. The principles on which CEN and ETSI DSRC tolling standards are based are quite similar, and their security algorithms are almost identical, so, considering the total number of on-board units deployed in Europe, it is worthwhile to spend a few words to describe them.

The tolling application is strictly master/slave, where the RSE is the master that initiates a transaction by issuing a broadcast message repeatedly. When an arriving OBE gets this broadcast message, it answers with some data, among which is its identifier. From that point onwards, a dedicated communication is established between the RSE and the OBE*. After that, data are read or written by using application commands. The data that are read or written in a given transaction are decided by the RSE, i.e., by each single toll charger, taken from a set that has to be available in each OBE. For example, the complete set of available data may include vehicle characteristics such as the plate number or the weight or dimensions of the vehicle, that may not be needed in a given tolling system that gets the same data by other means (cameras, or inductive loops). This in fact means that the tolling transaction is not unique, nor is it standardized in terms of sequence of issued commands and read/written data. An essential information, which is stored in all OBEs, is a reference to the contract with the service provider, by which the toll charger is able to associate the toll with an invoice. The same information is, in almost all cases, used in the security algorithms. Speaking of security, the communication channel is not secured, i.e., data are transmitted unencrypted. However, security measures are taken to ensure the identity of either the RSE or the OBE, or both. These security measures consist of exchanging authenticators that are computed by the RSE and the OBE by means of a known algorithm (specified in the standard), which operates on the basis of some data available in the OBE, randomly generated numbers, and the choice of an encryption key. As keys are symmetrical (use of asymmetrical keys requires more computing time, and time is essential in a DSRC communication), an

---

* The ways to ensure that the communication channel remains dedicated to the couple of RSE and OBE that are engaged in the dialogue may vary according to the different standards.

infrastructure to distribute master keys is needed, which is not defined in the standard, as well as an algorithm to compute derived keys is needed, which is specified in the standard instead.

6. Information Exchanges between Toll Chargers and Service Providers

Toll chargers and service providers need to exchange information on a number of subjects that are summarized by Fig. 9.

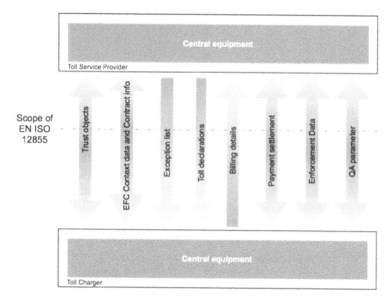

Figure 9: Data exchanges between toll chargers and service providers.

Information exchanged belong to different categories:

1. Security related information (trust objects), such as keys, certificates, certificate revocation lists, and so on.
2. Setup related information, such as the description of the toll system layout, or contractual details.
3. Tolling related information, such as toll declarations from the service provider in an autonoumous tolling system, or billing details from the toll charger, that contain all recorded data associated with a given user in a given path in a given tolling domain, to compute the invoices ultimately.
4. Payment settlement information, i.e., any aggregated set of data that will be used for actual invoices, or that confirm a payment to be done.

5. Enforcement data from the toll charger that show abnormal behaviour of users, or exception lists from a service provider, that indicate users to be treated in a special way (e.g. black lists to exclude some users, or white lists to indicate authorized users).
6. Quality data.

The data types and the protocol data units are defined in the standard (ISO 12855, see [12]), which, however, does not either specify the way data are physically transferred (the transfer protocol), nor impose Protocol Data Units to be mandatorily supported. Thus, a standard profile is needed to allow interoperability between applications, which, at the moment this paper is written, is being developed by CEN for the European environment.

7. EFC and ITS

One can raise the question of whether Electronic Fee Collection (EFC) belongs to the realm of Intelligent Transport Systems (ITS) or it is in fact a sector per se. On the one hand, it is true that, if ITS basically means "Information Technology applied to Transportation", EFC falls within the purview of ITS. On the other hand, however, the fact that EFC was born much before what is now considered ITS led to the development of a model where the application had to take into account all aspects that ITS applications now ignore, because they are taken care of by the underlying "system".

In other words, a typical ITS application relies on a set of services offered by the facility layer, ignoring how its data are transferred and it is possibly managed by a distributed standardized entity.

In contrast, EFC is an application that knows perfectly the technological infrastructure in which it operates and takes the maximum possible advantage from the features of the specific technology that is deployed. So from a puristic ITS point of view, EFC is one of those abhorred "silo applications", meaning applications that specify the whole stack of protocols and services by their own and ignore any structured architectural approach.

Yet, the principles of the tolling application as defined by the EFC standards are implementable as a "pure" ITS application, and this has been done, in some way, and described in a technical report recently published by CEN (CEN/TR 16690 in [20]). With time, the whole set of functionalities of the EFC set of standards will be mapped onto ITS station concepts by means of technical reports or other full standards, so that EFC as a whole will be undistinguishable from any other C-ITS application.

In conclusion: is EFC an ITS application? The question, from the view of the writer of this chapter, is completely irrelevant.

References

1. ISO 17573. 2010. Electronic fee collection – System architecture for vehicle-related tolling.
2. European Commission: Guide for the application of the Directive on the interoperability of Electronic Road Toll Systems. 2011. Publications Office of the European Union. ISBN 978-92-79-18637-0.
3. Malbrunot, François. 2006. The Long Term Cost of GPS/GSM vs. DSRC: How Do They Compare? In: *Tollways*, issue Winter. IBTTA.
4. ISO/IEC 18000. Information technology – Radio frequency identification for item management. Parts 1 to 7.
5. ISO 17264. Intelligent transport systems – Automatic vehicle and equipment identification – Interfaces.
6. EN 12834. Road transport and traffic telematics. Dedicated Short Range Communication (DSRC). DSRC application layer.
7. ISO 14906. 2011. Electronic fee collection – Application interface definition for dedicated short-range communication.
8. Jordán, Juan Guillermo, Francisco Soriano, David Graullera and Gregorio Martín. 2001. A comparison of different technologies for EFC and other ITS applications. Proceedings of IEEE Intelligent Transportation Systems Conference.
9. Dias, João R., S.R. Arnaldo, Oliveira and João Nuno Matos. 2013. The Charge Collector System: A New NFC and Smartphone-based Toll Collection System. Proceedings of Vehicular: The Second International Conference on Advances in Vehicular Systems, Technologies and Applications.
10. ISO/IEC 10746-1. 1998. Information Technology – Open Distributed Processing – Reference Model: Overview.
11. ISO 17575 Parts 1-3. 2015. Electronic fee collection – Application interface definition for autonomous systems.
12. ISO 12855. 2015. Electronic fee collection – Information exchange between service provision and toll charging.
13. EN 15509. 2014. Electronic fee collection - Interoperability application profile for DSRC
14. ETSI ES 200 674-1. 2013: Intelligent Transport Systems (ITS); Road Transport and Traffic Telematics (RTTT); Dedicated Short Range Communications (DSRC); Part 1: Technical characteristics and test methods for High Data Rate (HDR) data transmission equipment operating in the 5,8 GHz Industrial, Scientific and Medical (ISM) band.
15. ISO 12813. 2015. Electronic fee collection – Compliance check communication for autonomous systems.
16. ISO 13141. 2015. Electronic fee collection – Localisation augmentation communication for autonomous systems.
17. Association of Radio Industries and Businesses (ARIB). 2001. ARIB STD-T75 Dedicated Short-Range Communication System Version 1.0.
18. Association of Radio Industries and Businesses (ARIB). 2012. ARIB STD-T110 Dedicated Short-Range Communication System (Dsrc) Basic Application Interface Version 1.0.

19. CEN/TR 16690. 2014. Electronic fee collection – Guidelines for EFC applications based on in-vehicle ITS stations.

20. ITU-T Rec. F.400/X.400 | ISO/IEC 10021-1. 1999. Message handling system and service overview.

21. ISO/IEC 7498-1. 1994. Information Technology – Open Systems Interconnection – Basic Reference Model: The Basic Model.

Cooperative ITS: The SDO Perspective for Early Deployment

Hans-Joachim Fischer*

Elektrische Signalverarbeitung Dr. Fischer GmbH
Fichtenweg 9
D-89143 Blaubeuren, Germany

Abstract

"Cooperative Intelligent Transport Systems" (C-ITS) identifies a new paradigm in ITS that is reflected in several standards from various Standard Development Organizations (SDOs) intended to be used for day-1 deployment for the provision of road safety and traffic efficiency services.

Joint approaches of SDOs (CEN/ISO) and competitions between SDOs and groups of stakeholders (car makers, infrastructure operators, cellular network operators), together with international governmental attempts to achieve harmonized solutions, are explained. A short introduction in day-1 deployment use-cases is given. A major focus is on the ITS station architecture (ISO 21217) defining a station as a bounded secured and managed trusted device enabling essential features of the cooperative view – not just applicable for the safety applications context. Related standards being either necessary or beneficial to realize ITS station units and to test conformance, are introduced.

A discussion of the global context of road safety and traffic efficiency is followed by conclusions on required political decisions and a proposal for a "C-ITS Alliance" group being responsible for deployment issues, similar to the WiFi Alliance.

*Corresponding author: HJFischer@fischer-tech.eu

1. Introduction

Whilst Intelligent Transport Systems (ITS) is a work title in ISO since the 1990s covering a large number of topics of intelligent transport, see currently active Working Groups (WGs) from ISO and CEN in Fig. 1, the cooperative aspect of it and its usage to improve road safety by means of communications between users of the road network (cars, trucks, cyclists, pedestrians, etc.) was being standardized in various SDOs only since mid-2000.

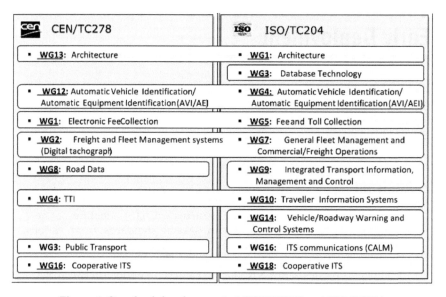

Figure 1: Standard development at CEN TC278 and ISO TC204
(Vienna Agreement on joint developments).

The term Cooperative ITS (C-ITS) was introduced by the CVIS project [61], and then adopted in 2009 as work title for the European Commission's mandate M/453 [50], which resulted in C-ITS standard developments at CEN TC278 and ETSI TC ITS, where CEN TC278 and ISO TC204 work jointly together under the Vienna Agreement illustrated in Fig. 1 (there are no similar agreements between CEN/ISO and other SDOs). The term "Cooperative ITS" (C-ITS) technically points to a feature of ITS, i.e. the sharing of data between different ITS applications, but also the sharing of communication resources by different application areas.

Due to independent standardization in various regions, performed by governmental SDOs and NGOs, see Fig. 2, the need for harmonization became obvious.

Figure 2: Some standard development organizations being active in C-ITS

In November 2009, EC/DGINFSO and USDOT/RITA signed a Joint Declaration of Intent on Research Cooperation in Cooperative Systems [51], which finally lead to the EU-US Cooperative Systems Standards Harmonization Action Plan (HAP) [52] in 2011. Work of this cooperation is performed by Harmonization Task Groups (HTGs), see Fig. 3.

Figure 3: Harmonization task groups.

The joint EU/US standardization approach linked CEN, ETSI, IEEE, ISO and SAE together.

- HTG 2 aimed at harmonizing Cooperative Awareness Message (CAM) [27], Decentralized Environmental Notification Message (DENM) [28], and Basic Safety Message (BSM) [29].
- HTG 1 and HTG 3 evaluated existing standards on communications and security and provided recommendations to SDOs on missing standards and standards to be harmonized. The deliverables of these two HTGs [53] are online available. Harmonization of messaging and service advertisement message formats ([30], IEEE WAVE [56] and ISO FAST [20, 31]) progressed well.
- For HTG 4 and HTG 5, the attempt was to achieve harmonized solutions for message sets, e.g. Signal Phase and Timing (SPaT) [25], directly from the beginning of development.
- HTG 6 works mainly on ITS Security Policies.
- In the meanwhile HTGs up to number nine are identified.

Although road safety is achieved mainly by appropriate design of vehicles, road network and autonomous systems at the roadside and inside vehicles such as (variable) message signs, sensor systems and radar, European car makers organized in the Car-to-Car Communications Consortium [65] developed at ETSI TC ITS use cases for road safety and traffic efficiency [6, 5] where communication between so-called ITS station units (ITS-SUs) aims at improving road safety and traffic efficiency. ETSI work focuses on communications at 5.9 GHz between vehicular ITS station units. CEN/ISO worked on further use cases with a focus on infrastructure needs, i.e. requiring communications between vehicular ITS-SUs and fixed ITS-SUs (at the roadside and in traffic control centers) based on a multitude of access technologies (hybrid communications).

The concept and the architecture of an ITS-SU and related communications was developed at ISO TC204 (ISO 21217 [4]), and later adopted by CEN TC278 and by ETSI TC ITS. ETSI TC ITS presented this architecture in EN 302 665 [58]; this duplication of standard obviously caused some confusion; however, the technical content is essentially identical.

2. Use Cases for Road Safety and Traffic Efficiency

A large number of use cases for C-ITS is already developed in various SDOs and projects around the globe. For day-1 deployment in Europe, e.g. in the EU C-ITS corridor from Rotterdam in The Netherlands via Frankfurt in Germany [62] to Vienna in Austria [63], infrastructure-related use cases were selected:

- Road Works Warning (RWW) [54, 62]
  Road Works Warning informs drivers of road works ahead, their relevant parameters and associated obstructions (e.g. closed lanes). The purpose is to alert the driver in time to increase awareness and to inform of potentially dangerous conditions.
- Improved traffic management by incorporating Probe Vehicle Data (PVD) [53, 61]
  The collection of anonymized information from vehicular ITS stations enlarges the information basis for traffic management decisions.
- In-Vehicle Information (IVI) [54]
  In-Vehicle Information standardized in ISO 19321 [26] is used to inform drivers about present speed policy/advice and other relevant (hazard) information which are shown on dynamic and/or static traffic signs.
- Intersection Safety (ISS) [54]
  Cooperative Traffic Lights provide information on the traffic semaphore status (SPaT – Signal Phase and Timing standardized in ISO 19091 [25]) and a geographical representation of the vicinity of the traffic light (MAP).
- and others.

The German CONVERGE project [60], focusing on the development of business models and appropriate system architecture, successfully implemented, validated and demonstrated e.g. the vehicle-related use case.

- Wrong Way Driver Warning
  Wrong Way Driver Warning indicates to vehicles in the affected area that a vehicle is driving against the planned direction of traffic. The affected area is primarily the road in which the vehicle is going in the wrong direction and the affected vehicles are those approaching the violating vehicle.

These use cases require communication between vehicular ITS-SUs and fixed ITS-SUs (roadside units and traffic control centers). Two complementary wireless communication protocol stacks (this is the uppermost simple hybrid communications approach) are necessary, not just beneficial, i.e.

- Single-hop communications over a short range narrow-band channel at 5,9 GHz;
- Cellular network communications;

both optionally or alternatively using the Internet protocol IPv6 with mobility support. Further on, Internet connectivity and private legacy technologies are used for wired communications.

2.1 Communications in C-ITS

The ITS station (ITS-S) and communications concept was originally developed at ISO (ISO 21217 [4]), see the simplified ITS-S architecture in Fig. 4.

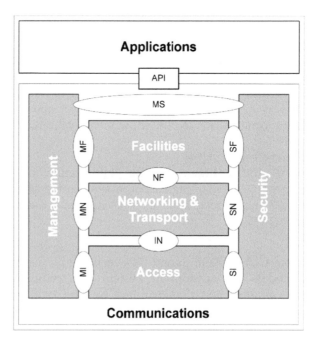

Figure 4: ITS-S simplified architecture [4].

The term ITS-S defines a set of functionality, e.g. the following:

- Operation (communications and services) in a Bounded Secured Managed Domain (BSMD).
- Support of multiple communication protocol stacks for wired and wireless connectivity described with a simplified OSI model consisting of three combined layers:
  o ITS-S access layer: OSI 1 and 2;
  o ITS-S networking & transport layer: OSI 3 and 4;
  o ITS-S facilities layer: OSI 5, 6 and 7.
- Station management and security at the side of the communication protocol stack:
  o ITS-S management entity;
  o ITS-S security entity.

- Multiple applications providing ITS services to the user:
  - ITS-S applications: on top of ITS-S facilities layer, the ITS-S management entity, and the ITS-S security entity.

This architecture supports any kind of implementation (single box, multiple boxes) for a variety of usage contexts, see Fig. 5, e.g.:

- mobile devices (device installed in a vehicle, a portable device)
- fixed devices (device installed at the roadside, device installed in a traffic control center).

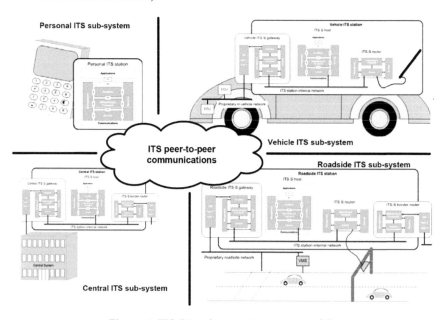

Figure 5: ITS-S implementation contexts [4].

Important features are:

- Abstraction of applications from communications which makes this design scalable and future-proof. That means that an application does not need to request usage of a specific communications protocol stack for transmission of Application Protocol Data Units (APDUs), but just presents functional requirements for communications to the station management. The station management will then match the requirements dynamically with available capabilities and select the most appropriate communications protocol stack.
- Efficient handover methods between different kinds of access technologies and communication protocols. This means that communication channels, access technologies and even

networking & transport layer protocols may change dynamically in order to provide an optimum overall performance.

- Multiple methods for sharing of data between different ITS-S applications:
  - o Sharing of messages and parts of it (data) between different applications using real-time "publish-subscribe mechanisms".
  - o Sharing of geo-referenced and time-referenced data stored in a Local Dynamic Map (LDM) (such data being produced typically by sensors in vehicles and at the roadside are referred to as Probe Data).

- Secured interfaces to non-secured devices, e.g.
  - o gateway to the CAN bus of a vehicle to access Vehicle Probe Data;
  - o gateway to portable devices granting them access to a default communication channel for Internet access.

Applications implementing road safety related use-cases impose stringent requirements on communications, such that communication protocols with very low latency, very limited protocol overhead, prioritized access and security means (authentication/encryption) are needed.

Methods to disseminate information in a timely manner at specific locations identified by geo-coordinates as identified in the EC GeoNet project [59] enables customized support of road users [57]. Deployment projects such as the German CONVERGE project [60] clearly expressed the need for general geo-dissemination protocols being applicable for any kind of communication protocol stack.

Channel congestion is a major problem, especially in typical road traffic scenarios on multi-lane roads and during rush hour. In order to ensure a sufficient level of road safety enhancement by C-ITS, the station management uses channel congestion control algorithms including support of different access technologies and different communication protocols. Thus for early deployment of C-ITS, an efficient and flexible general approach was identified in various projects:

- Usage of at least two access technologies, i.e. (1) a short range (primarily single-hop) access for ad-hoc communications between peer entities at 5.9 GHz, and (2) cellular network access (e.g. LTE).
- Usage of (a) highly optimized non-networking messaging and service advertisement protocols, and (b) IPv6 networking.

3. Base Standards

The main purpose of standardization is to enable interoperability of equipment provided by different vendors. This covers e.g.:

- specification of exposed interfaces over which protocol data units are exchanged in a testable way
- data and message specifications
- specification of protocol behavior including handling of unexpected events.

However, standards should never be descriptive system specifications but allow for options. Usage of options is selected in profile standards as used in system specifications.

To assist vendors to implement features, standards may also specify station-internal features such as station management, which becomes at least beneficial – not to say "a prerequisite" – once remote management is added to the system specification, or for distributed implementations of an ITS-SU with several physical units provided by different manufacturers.

Base standards are only part of the play; these need to be complemented by standardized conformance test suites.

The following base standards are in support of safety-critical C-ITS applications. Most of them are ready for use in early deployments, referred to as "Release 1". The column "Reference" in the subsequent tables contains

- in brackets a reference number extending the numbering in chapter "References", and
- the SDO document reference number.

3.1 Architecture Standards, Use Cases, and Tutorials

Reference	Title	Comment
[1] ISO 17419	Intelligent transport systems – cooperative systems – classification and management of ITS applications in a global context	To a large extent a tutorial on trust. Specifies format of globally unique identifiers.
[2] ISO 17427	Intelligent transport systems – cooperative systems – roles and responsibilities in the context of cooperative ITS based on architecture(s) for cooperative systems	Architecture of C-ITS.
[3] ISO 17427-x	Intelligent transport systems – cooperative ITS – Part 1: < this will replace ISO 17427 > Part 2: Framework overview Part 3: Concept of operations (ConOps) for 'core' systems	Set of tutorial standards on C-ITS (Technical reports)

Reference	Title	Comment
	Part 4: Minimum system requirements and behavior for core systems Part 5: Common approaches to security Part 6: 'Core system' risk assessment methodology Part 7: Privacy aspects Part 8: Liability aspects Part 9: Compliance and enforcement aspects Part 10: Driver distraction and information display Part 12: Release processes Part 13: Use case test cases Part 14: Maintenance requirements and processes	
[4] ISO 21217	Intelligent transport systems – communications access for land mobiles (CALM) – architecture	ITS station and communications architecture.
[5] ETSI TR 102 638	Intelligent Transport Systems (ITS); vehicular communications; use cases; description	C-ITS road safety and traffic efficiency use cases.
[6] ETSI TS 102 637-1	Intelligent Transport Systems (ITS); vehicular communications; basic set of applications; part 1: functional requirements	
[7] ISO 20025	Intelligent transport systems – cooperative ITS – representative probe data use cases and related gaps in existing probe data standards	Technical report on early investigations in probe data and related standardization needs.

3.2 Station Management Standards

Reference	Title	Comment
[8] ISO 17423	Intelligent transport systems – cooperative systems – ITS application requirements and objectives for selection of communication profiles	In support of abstraction of applications from communications.

Reference	Title	Comment
[9] ISO 24102-1	Intelligent transport systems – communications access for land mobiles (CALM) – ITS station management – part 1: local management	Local station management.
[10] ISO 24102-2	Intelligent transport systems – communications access for land mobiles (CALM) – ITS station management – part 2: remote management of ITS-SCUs	Remote station management.
[11] ISO 24102-3	Intelligent transport systems – communications access for land mobiles (CALM) – ITS station management – part 3: service access points	Station-internal service access points MI, MN, MF, MS, SI, SN, SF, and MA, SA contained in API, see Fig. 4.
[12] ISO 24102-4	Intelligent transport systems – communications access for land mobiles (CALM) – ITS station management – part 4: station-internal management communications	In support of distributed implementation of ITS-SUs.
[13] ISO 24102-6	Intelligent transport systems – communications access for land mobiles (CALM) – ITS station management – part 6: path and flow management	In support of abstraction of applications from communications.

3.3 Application Support Standards

Reference	Title	Comment
[14] ISO 17429	Intelligent transport systems – cooperative ITS – ITS station facilities for the transfer of information between ITS stations	Facilities layer services in support of sharing messages and message data between different applications. Also specifies a common message handling approach.
[15] ISO 18750	Intelligent transport systems – cooperative ITS – local dynamic maps	Facilities layer service in support of sharing geo-referenced and time-stamped data between different applications. Station-internal data store.

Reference	Title	Comment
[16] ISO 21176	Intelligent transport systems – cooperative ITS – position, velocity and time functionality in the ITS station	Facilities layer service providing reliable and accurate position, velocity and time information.
[17] ISO 21177	Intelligent transport systems – secure vehicle interface – ITS-station security services for secure session establishment and authentication	New work items related to an in-vehicle gateway allowing access to probe data.
[18] ISO 21184	Intelligent transport systems – secure vehicle interface – data dictionary of vehicle-based information for C-ITS applications	
[19] ISO 21185	Intelligent transport systems – secure vehicle interface – communication profiles for secure connection between an ITS-station and a vehicle	
[20] ISO 24102-5	Intelligent transport systems – communications access for land mobiles (CALM) – ITS station management – part 5: Fast Service Advertisement Protocol (FSAP)	Message format specified by reference to ISO 16460. Interoperability mode with IEEE WSA [55].

3.4 Application and Data Definition Standards

Reference	Title	Comment
[21] ISO 17426	Intelligent transport systems – cooperative systems – contextual speeds	About speed limit information presented inside a vehicle.
[22] ETSI TS 101 539-1	Intelligent Transport Systems (ITS); V2X applications; part 1: Road Hazard Signalling (RHS) application requirements specification	Road safety support applications

Reference	Title	Comment
[23] ETSI TS 101 539-3	Intelligent Transport Systems (ITS); V2X applications; part 3: Longitudinal Collision Risk Warning (LCRW) application requirements specification	

3.5 Message and Data Standards

Reference	Title	Comment
[24] ISO 17425	Intelligent transport systems – cooperative systems – data exchange specification for in-vehicle presentation of external road and traffic related data	Special case of in-vehicle information [26] with extended features
[25] ISO 19091	Intelligent transport systems – cooperative systems – using V2I and I2V communications for applications related to signalized intersections	Signal Phase and Timing (SPaT)
[26] ISO 19321	Intelligent transport systems – cooperative ITS – dictionary of in-vehicle information (IVI) data structures	Data dictionary (also covers contextual speed data [24])
[27] ETSI EN 302 637-2	Intelligent Transport Systems (ITS); vehicular communications; basic set of applications; part 2: specification of cooperative awareness basic service	Periodically transmitted cooperative awareness message. Corresponds with Basic Safety Message in USA [29].
[28] ETSI EN 302 637-3	Intelligent Transport Systems (ITS); vehicular communications; basic set of applications; part 3: specifications of decentralized environmental notification basic service	Notification of events.
[29] SAE J2735	Dedicated Short Range Communications (DSRC) message set dictionary	Periodically transmitted basic safety message. Corresponds to cooperative awareness message in EU [27].

3.6 Communication Protocol Standards

Reference	Title	Comment
[30] ISO 16460	Intelligent transport systems – Communications Access for Land Mobiles (CALM) – communications protocol messages for global usage	Message formats harmonized between ISO TC204 WG16 (CALM FAST) [20, 31] and IEEE 1609 WG (IEEE WAVE) [56].
[31] ISO 29281-1	Intelligent transport systems – Communication access for land mobiles (CALM) – non-IP networking – part 1: Fast Networking & Transport layer Protocol (FNTP)	Efficient networking & transport layer protocol for non-IP communications. Message format specified by reference to ISO 16460 [30]. Interoperability mode with IEEE WSMP [56].
[32] ISO 21210	Intelligent transport systems – Communications access for land mobiles (CALM) – IPv6 networking	Enables mobile stations in IPv6.
[33] ISO 16788	Intelligent transport systems – Communications access for land mobiles (CALM) – IPv6 networking security	
[34] ISO 16789	Intelligent transport systems – Communications access for land mobiles (CALM) – IPv6 networking optimization	

3.7 Access Technology Standards

Reference	Title	Comment
[35] ISO 17515-1	Intelligent transport systems – Communications access for land mobiles (CALM) – Evolved Universal Terrestrial Radio Access Network (E-UTRAN) – part 1: general usage	LTE network access.

Reference	Title	Comment
[36] ISO 17515-2	Intelligent transport systems – Communications access for land mobiles (CALM) – Evolved Universal Terrestrial Radio Access Network (E-UTRAN) – part 2: device to device communications (D2D)	New work item on LTE device to device communications.
[37] ISO 21213	Intelligent transport systems – Communications access for land mobiles (CALM) – 3G cellular systems	3G cellular network access.
[38] ISO 21214	Intelligent transport systems – Communications access for land mobiles (CALM) – infra-red systems	Infrared communications. Suited for bumper-to-bumper communications. Currently not considered for deployment in C-ITS.
[39] ISO 21215	Intelligent transport systems – Communications access for land mobiles (CALM) – M5	Based on IEEE 802.11 (p-mode) [55]. Frequency allocation for C-ITS at 5.9 GHz.
[40] ISO 21218	Intelligent transport systems – Communications access for land mobiles (CALM) – access technology support	Enables usage of legacy technologies.
[41] ETSI TS 102 687	Intelligent Transport Systems (ITS); decentralized congestion control mechanisms for intelligent transport systems operating in the 5 GHz range; access layer part	European approach for congestion control at the ITS-S access layer.
[42] ETSI TS 102 724	Intelligent Transport Systems (ITS); harmonized channel specifications for intelligent transport systems operating in the 5 GHz frequency band	European usage of 5,9 GHz frequency bands

Security standards

Security standards from various SDOs exist. Usage of them is subject to ongoing discussions with authorities. Major issues are:

- protection of humans compliant with regulations on privacy;

- authentication of senders of messages;
- establishing trusted devices.

Obviously there are different regulations in various regions of the world, but also in countries of the same region. That makes an interoperable approach quite difficult. A breakthrough cannot be achieved at the technical level of SDOs, but only at the administrative level of national authorities.

The following table just shows some examples of security standards, not covering all issues:

Reference	Title	Comment
[43] IEEE 1609.2™	IEEE standard for Wireless Access in Vehicular Environments (WAVE) – security services for applications and management messages	The US view on security details for WAVE.
[44] ETSI TS 102 731	Intelligent Transport Systems (ITS); security; security services and architecture	The ETSI view on security architecture.
[45] ETSI TS 102 940	Intelligent Transport Systems (ITS); security; ITS communications security architecture and security management	The ETSI view on communications security architecture and security management.
[46] ETSI TS 102 867	Intelligent Transport Systems (ITS); security; stage 3 mapping for IEEE 1609.2	The ETSI view on how to use IEEE 1609.2 in Europe.

3.8 Test Standards

Reference	Title	Comment
[47] ISO/IEC 9646 (all parts)	Information technology – open systems interconnection – conformance testing methodology and framework	The "root" standard on conformance testing.
[48] ETSI EG 202 798	Intelligent Transport Systems (ITS); testing; framework for conformance and interoperability testing	General guidelines on testing in C-ITS. Based on [47]. This is complemented by [49].

Reference	Title	Comment
[49] ISO 20026	Intelligent transport systems – cooperative ITS – test architecture	Test architecture for ITS-SUs implementing ISO 21217 [4] and ISO 24102-4 [12]. This approach reduces test expenses by a standardized test interface. Complements [48].
ETSI and CEN/ISO	Conformance test suites for many base standards	Details not presented here.

4. Testing

A base standard cannot be considered to be ready for usage before at least the related conformance test standard is validated. In a second step, interoperability testing should also be performed. Whilst conformance testing checks protocol details of a single implementation, interoperability testing is a kind of black box testing of implementations from different vendors. An introduction to testing in C-ITS is provided in [48].

In C-ITS, conformance test standards are already available for almost all base standards. The ETSI Centre for Testing and Interoperability (CTI) continuously invites for C-ITS plug-tests and interoperability trials. Such a conformance test standard consists of three parts:

1. Protocol Implementation Conformance Statement proforma (PICS)
2. Test Suite Structure and Test Purposes (TSS&TP)
3. Abstract Test Suite (ATS).

PICS is a template that allows an applicant to declare how his implementation conforms with a standard. All testable features can be selected to be supported or not supported, and dynamic features (e.g. supported values of parameters) can be declared.

TSS&TP describes details of the test setup consisting of a "Test System" (TS) and the "Implementation under Test" (IuT) contained in a "System under Test" (SuT). Further on, it specifies tests in human readable language.

ATS provides machine-readable TTCN-3 code to execute conformance tests.

5. The Global Context of C-ITS

As presented above, road safety is mainly achieved with autonomous systems as only this approach is (1) efficient, and (2) does not bear unpredictable legal (liability) risks. Communications can only add a bit to the safety level achieved with autonomous systems, especially by providing information to road users that is not very time critical. And – there is no business model for road safety.

A bigger part of C-ITS is the domain of traffic efficiency. A key is to know better about the real traffic situation, which can be achieved by getting access to probe data, especially probe vehicle data (PVD). Road operators want to use probe data to improve traffic management. These data are needed (1) directly from the vehicles (raw data) at selected positions in the road network, and (2) from "the cloud" that pre-process raw data from a large number of vehicles. Access to vehicle probe data currently is heavily discussed. Car makers claim to own the PVD. They want to send PVD to their own cloud, pre-process the data, and sell the processed data to the road authorities and road users. This view of the car makers obviously is not shared by car owners and by road authorities. Another discussion is on privacy related to PVD. Latest understanding is, e.g. expressed by Mr. Sigmar Gabriel, Minister of the German Federal Ministry of Economy and Energy, at the carIT congress [70], to replace privacy by sovereignty. That means, the one who controls data can (ignore privacy, and) give the data to whom he wants in a sovereign way (in line with the customers needs and wishes). These discussions are mirrored in debates at e.g. ISO, where TC204 wants standardizing an in-vehicle gateway as part of an ITS-SU to access PVD in a secure way, and TC22 wants to standardize cloud-related features.

Further bigger part of C-ITS are the "social networking domain" and the "commercial domain". The commercial domain is prepared in a way so as to receive service advertisements, and consume the announced services either as a result of a communication session, or by taking the received advice on where the service is provided.

As C-ITS is not another new silo technology, but actually breaks the "silo-thinking", an ITS-SU manages quite a different service domains with different societal values, and thus different priorities. Consequently the approach in ISO 21217 [4] of an ITS-S as a BSMD is a pre-requisite for successful deployment of the whole C-ITS spectrum in a single device.

6. Conclusions

The approach in C-ITS standardization breaking the conventional "silo-thinking" of systems seems to be the biggest problem to introduce C-ITS quickly on the roads and in the road infrastructure. A severe conflict is presented by the fact that three big stakeholders finally "meet in a single device" installed in vehicles. These stakeholders are the car makers, the road operators and city authorities, and the cellular network operators. They have different business models and different use cases.

Without governmental "support" requesting the introduction of C-ITS with defined features, a proliferation of the market might result in a situation having two or more C-ITS devices in a vehicle, i.e. a road safety ITS-SU installed by the car maker, an after-sales general purpose ITS-SU with no access to the PVD, a navigation device supporting some C-ITS services, handheld smart phones supporting some C-ITS services.

The issue of trust and security requires availability of a Public Key Infrastructure (PKI) which has to be provided most likely by a governmental trust authority.

To promote early deployment of C-ITS, a European C-ITS corridor project was launched by the Governments of The Netherlands, Germany [62], and Austria [63]. It became obvious that this project to a certain extent also suffers from problems presented above.

What is finally needed to run the C-ITS is a kind of globally active "C-ITS alliance" with regional pillars that constitutes the C-ITS authority that is needed to ensure proper operation in a trusted way. So far only regional attempts are existent such as the Amsterdam Group [64] in Europe that is unfortunately not yet fairly well balanced although car makers [65] and road operators [66, 67] and city authorities [68] are members. It is noted that luckily the World Road Association PIARC [69] is working intensively on C-ITS, processing information from ISO TC204.

References

References 1 to 49 are in the tables in the section on "Base Standards".
50. Standardisation Mandate Addressed to CEN, CENELEC and ETSI in the field of information and communication technologies to support the interoperability of co-operative systems for intelligent transport in the European Community. European Commission, Enterprise and Industrial Directorate-General, Brussels, 6. October 2009.
51. EU-US Joint Declaration of Intent on Research Cooperation in Cooperative Systems. 2009. EC/DGINFSO and USDOT/RITA.
52. EU-US Cooperative Systems Standards Harmonization Action Plan (HAP), EC DG INFSO and USDOT RITA JPO. 2011.
53. Feedback to ITS Standards Development Organizations – Communications, EU-US ITS Task Force – Standards Harmonization Working Group - Harmonization Task Group 3, Document HTG3-3, http://ec.europa.eu/digital-agenda/en/news/progress-and-findings-harmonisation-eu-us-security-and-communications-standards-field.
54. EU C-ITS Corridor, Austrian project ECo-AT, SWP 2.1 Use Cases – Overview on Use Cases. 2015. Version 03.00.
55. IEEE 802.11™. 2010. IEEE Standard for Information technology – Telecommunications and information exchange between systems Local and metropolitan area networks – Specific requirements Part 11, Wireless LAN Medium Access Control (MAC) and Physical Layer (PHY) Specifications.

56. IEEE 1609.3™ 2016. IEEE Standard for Wireless Access in Vehicular Environments (WAVE) – Networking Services.
57. ETSI EN 302 636-4-1, Intelligent Transport Systems (ITS); Vehicular Communications; GeoNetworking; Part 4: Geographical addressing and forwarding for point-to-point and point-to-multipoint communications; Sub-part 1: Media-Independent Functionality.
58. ETSI EN 302 665 V1.1.1. 2010. Intelligent Transport Systems (ITS); Communications Architecture.
59. GeoNet project, http://www.geonetproject.eu/
60. CONVERGE project, http://www.converge-online.de/
61. CVIS project, http://www.cvisproject.org/
62. EU C-ITS Corridor, German project web site: http://c-its-corridor.de/?menuId=1&sp=en
63. EU C-ITS Corridor, Austrian project web site: http://eco-at.info/
64. Amsterdam Group, http://www.amsterdamgroup.eu/
65. Car-to-Car Communications Consortium, https://www.car-2-car.org/index.php?id=5
66. ASECAP, http://www.asecap.com/index.php?lang=en
67. CEDR, http://www.cedr.fr/home/
68. POLIS, http://www.polis-online.org/
69. PIARC, http://www.piarc.org/en/
70. carIT congress. 2015. IAA Frankfurt, Congress Center Messe, Frankfurt.

Probe Vehicle Information Systems

Masaaki Sato[1]*, Manabu Tsukada[2] and Hiroshi Ito[3]

[1]Graduate School of Media and Governance, Keio University,
Interactive and Digital Media Institute, National University of Singapore
[2]Graduate School of Information Science and Technology,
The University of Tokyo
[3]ITS Research Division, Japan Automobile Research Institute (JARI)

Abstract

The fusion of automobile technology and information technology is now progressing at a rapid pace. The use of information technology in automobiles is being developed in two ways. One is its role in enhancing driving performance, so that some cars have begun to resemble "computers that move". Another key area is Intelligent Transport Systems (ITS) that provide innovative services relating to transport, traffic management and can enable car users to be better-informed and smarter users of the technology. A probe vehicle system, which utilizes the sensor data of each automobile, has become a new trend in the service deployment of ITS for enhancing car-telematics. This kind of system is expected to provide a platform for large amounts of data based on automobiles. This section describes Probe Vehicle Systems that exist, and various methods of fusion between the Internet and automobile technologies. In addition, the possibilities of developing various interfaces and forms of privacy protection that can be integrated into probe vehicle systems, and which can also provide a basis for various services, are also explored and examined in depth. Moreover we mention some international standardization activities.

*Corresponding author: saikawa@sfc.wide.ad.jp

1. Introduction

Overview

In the age where the basis of Internet infrastructure for digital communication is developing, one of the most important issues that we need to address urgently is the building of a global service infrastructure on our society based on our activity and mobility.

The cutting edge network technologies are continuing to evolve at a remarkable speed and compelling environment in which we carry out various activities transparently and not limited to fixed locations. In such mobility aware environment, infrastructure and service based on mobility have become necessity for transparent and optimal activities. With such mobile activity focuses, collecting, sharing and representing valuable probe information from vehicular activities using the Internet is becoming a hot topic in the research field and deployments beyond many countries.

One example, in Asia the proliferation of wealthy and middle class people has accelerated urbanization and redevelopment and, thereby, has brought about a change in their lifestyle, such as motorization. Car markets in Asian countries have increased rapidly by collaboration with Japanese, European and U.S. automobile companies. On the other hand, various problems also occur with the increase in the number of vehicles, such as traffic congestion, accidents and environmental pollution. For example, economic loss by traffic congestion in Jakarta is 5.5 million Singapore dollars per year, as estimated by Indonesia government. They would like to solve the terrible problem utilizing digital information and communication technologies.

Intelligent transportation systems (ITS) are advanced applications aimed to provide innovative services related to different modes of transport and traffic management and enable various users to be better informed and make safer, more coordinated and 'smarter' use of transport networks. ITS have two major goals, one is a reduction of the number of fatalities and injuries on the roads and another is a reduction of hours of traffic congestion. The latter goal would also contribute to reducing the environmental impact of transport. In addition, the fusion of ITS and information technology (IT) is progressing at a rapid pace. ITS offer a fundamental solution to various issues concerning transportation, which include traffic accidents, congestion and environmental pollution. ITS deal with these issues through the most advanced communication and control technologies. ITS could be provided better communication for humans, roads and automobiles. By creating ideal traffic conditions, ITS system will reduce traffic accidents and congestion while saving energy and protecting the environment. Recently, Cooperative-ITS that is an information exchange and sharing platform utilizing V2X (Vehicle to Vehicle/Road-side Infrastructure, pedestrian, etc. ...) communication instead of V2I/I2V communication. Such a cooperative-ITS, the realization of the more efficient road traffic, CACC (Cooperative Adaptive Cruise Control) and autonomous driving is expected.

The Probe Vehicle System plays an increasing role in contributing to these top priorities of our society. Probe data is information from sensors of probe vehicle, such as location, speed, acceleration, etc. In earlier works also called floating car data (FCD) this information stems mainly from sensors and contains at least position information, but may also include vehicle traction information, driver actions (steering, braking), weather and road surface conditions, etc. Figure 1 presents the overview of Probe Vehicle System.

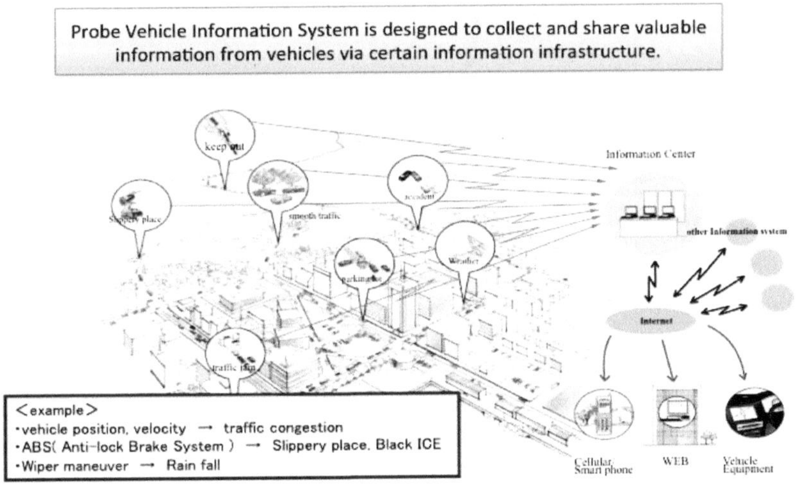

Figure 1: The overview of Probe Vehicle System [1].

Basically Probe Vehicle System is intended to provide traffic information, travel information, surrounding environmental information or information about road (surface) conditions. In addition, Probe Vehicle System is also effective in a disaster situation, not only in a normal one. In the disaster situation, many road side equipments would be destroyed and the power outage would occur here and there. Even in such a situation, Probe Vehicle System can be operated. In a terrible situation caused by Great East Japan Earthquake 11th March 2011, one could estimate road damage utilizing probe data. Active road map is one of the most important sources of information for vehicles of emergency service, rescue and other disaster response organizations. Figure 2 shows active road map made by probe data at Great East Japan Earthquake situation.

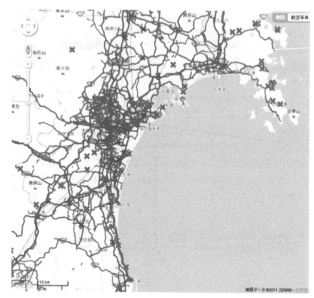

Figure 2: Active road map made by probe data. (From ITS Japan WEB page)

1.1 Definition

A vehicle has more than one hundred sensors. Useful information collected from these sensors can serve as a secure foundation for society. The probe vehicle system is collecting the vehicle sensor data via a communication infrastructure. The probe vehicle system is researched, studied or examined by many countries. Besides, the general architecture, common data format and interfaces were standardized as ISO 22837:2009 [2] in ISO/TC204/WG16.

The following list is denoted in the definition regarding the main components of probe vehicle system. The relationship between the factors is illustrated in Fig. 3.

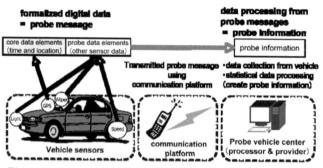

Figure 3: Concept and definition of probe vehicle system.

- Vehicle sensor:
  A device on a vehicle that senses conditions inside and/or outside the vehicle or that detects actions that the driver takes, such as turning on/off headlights or windshield wipers, applying the brakes, etc.
- Probe data:
  Vehicle sensor information that is processed, formatted, and transmitted to a land-based center for processing to create a good understanding of the driving environment. This probe data includes probe data elements and probe messages.
- Probe data element:
  An item of data included in a probe message, typically from onboard sensors. Systems in the vehicle may do some processing on the sensor reading to convert it into a suitable form for transmission.
- Core data element:
  Core data elements are basic descriptive elements intended to appear in every probe message. Core data element consists of a timestamp and a location stamp describing the time and place at which the vehicle sensor reading was made.
- Probe message:
  The result of transforming and formatting one or more probe data elements into a form suitable to be delivered to the onboard communication device for transmission to a land-based center. It is emphasized that a probe message should not contain any information that identifies the particular vehicle from which it originated or any of the vehicle's occupants, directly or indirectly.
- Processed probe data:
  The result of fusing and analyzing data from probe data messages in combination with other data.

On the other hand, The United States (US) Department of Transportation (USDOT) and the Road Bureau of Ministry of Land, Infrastructure, Transport, and Tourism (MLIT) of Japan have a long history of sharing information on ITS (Intelligent Transportation Systems) activities. A US-Japan ITS Task Force was established specifically to facilitate the exchange information and identify areas for collaborative research for the development and deployment of ITS in the US and Japan. The Task Force also created probe data concept and definition that form a public transportation management's point of view. The definition as follows [3]:

Probe data are data generated by vehicles (light, transit, and freight vehicles) about their current position, motion, and time stamp. Probe data also include additional data elements provided by vehicles that have added intelligence to detect traction information, brake status, hard braking, flat tire, activation of emergency lights, anti-lock brake status, air bag deployment status, windshield wiper status, etc. Probe data from vehicles may be generated by devices integrated with the vehicles' computers, or nomadic devices brought in to the vehicles.

Probe data does not include data that have been derived outside of the vehicle, even if these data were aggregated from data generated by vehicles. For example, travel times that are derived from position and motion data are not classified as probe data.

Probe data may be transmitted at various frequencies (i.e., every 10th of second, every 1 minute, whenever a vehicle enters a roadside Dedicated Short Range Communication (DSRC) communication area, whenever an event is triggered, etc.), using a range of wireless communication technologies, including DSRC, cellular, Wi-Fi, WiMAX, etc.

1.2 Reference Architecture

In ISO 22837:2009, the reference architecture for probe vehicle systems presents the initial categorization of system components and their relationships from a conceptual point of view. A component is depicted as a UML class and represents an encapsulation of functions and data that is conceptually considered an individual entity in the probe vehicle system. A relationship is depicted as a UML association and represents potential control and/or data flow among components. The reference architecture is indicated in Fig. 4.

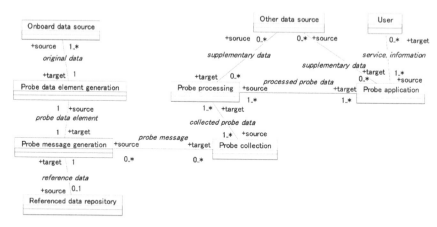

Figure 4: The overall structure of the reference architecture for probe vehicle systems. (From ISO 22837:2009)

The followings are the components of this reference architecture.

- Onboard data source:
  The onboard data source provides original data that will become a probe data element. Original data may be raw sensor data or data

from other onboard applications. Onboard data sources may be (various types of) sensors, onboard systems and so on.

- Probe data element generation:
  Probe data element generation creates probe data elements from original data. All of the following cases are included: (1) no processing (probe data element is identical to original data), (2) normalize original data (probe data element is the result of performing a calculation or transformation of original data) and (3) process original data to generate a new type of data (multiple items of original data are processed, possibly over a time period, to produce the probe data element, e.g., "traffic jam detected").
- Probe message generation:
  Probe message generation creates and formats probe messages from probe data elements and sends them to probe collection. Here, "send" is at the application layer, not the communication layer. Probe message generation manages the timing of sending messages as an application issue. Actual message transmission out of the vehicle is left to the communication layer. Probe message generation may refer stored reference data, to assist with data transformation or to help determine whether a probe message should be sent.
- Referenced data repository:
  Referenced data repository holds data for reference by the probe message generator.
- Probe collection:
  Probe collection is a land-side activity that receives probe messages sent by vehicles and extracts probe data from these messages.
- Probe processing:
  Probe processing receives collected probe data from probe collection and processes it (for example, using analysis and fusion). Probe processing does not receive any information from probe collection that identifies the vehicle or driver.
- Probe application:
  Application which uses information produced by probe processing.
- Other data source:
  Other data source provides additional data that is used for probe processing and/or by probe applications. Other data sources may be road authorities, police, weather information providers and so on.
- User:
  It is the entity that receives services and/or information produced from probe data. Users may be drivers, road authorities, police, weather services, public agencies, individual users (of cell phones, PDAs) and so on. Each relationship in this reference architecture is represented as a data and/or control flow. The followings are those data and flow in this reference architecture.

- Original data:
  This is data used for probe data generation. Original data may be raw sensor data or data from other onboard applications.
- Reference data:
  This is data stored in a repository and referred to for probe data generation. Reference data may be (among other things) historical data and/or statistical data.
- Collected probe data:
  This is probe data collected by the probe collection component, to be sent to probe processing components.
- Supplementary data:
  This is data from other data sources (non-vehicle) that is also used in probe processing and/or by probe applications.
- Processed probe data:
  This is data from probe data messages which has been collated and analyzed in combination with other data.
- Service, information:
  It is the value-added result of combining processed probe data with supplementary data for delivery to users.

1.3 Information Model

Basically, service provider can deliver two types of service utilizing probe vehicle systems. One is "snapshot" information, like a traffic congestion, accident point, slippery place, heavy raining and so on. Service receivers (users) know "what happens now" or "an event was there at that time" utilizing provided information based on probe data. If service providers have the capability of storing a lot of probe data, they can provide statistical traffic information (e.g. This road is often crowded on Sunday morning).

Another one is "Consecutive" information, like a multi link travel time, waiting time for a traffic signal, the cost estimation of turning right (or left). In this case, services providers need consecutive (linked) data group for services. The ability to provide these services is a big advantage for probe vehicle systems, because it is very difficult for other infrastructure systems to create the vehicle sensor data with linkability. A linkability of probe data sometimes causes privacy issues to probe data sender. Information model of two types of services is illustrated in Fig. 5.

1.4 Consideration of Probe Vehicle Systems

Probe vehicle systems have become a new trend for service deployment of ITS (Intelligent Transport Systems) to enhance Car Tele-matics. For measuring the vehicle density and the average speed, road operators have invested in the past in expensive infrastructure, inductive loops in the pavement, microwave, ultrasound and infra-red sensors, gantries

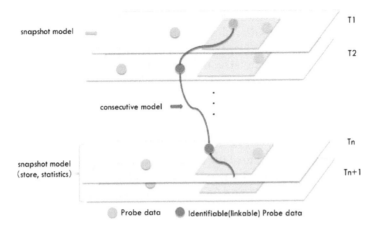

Figure 5: Information model of services utilizing probe data.

with video cameras, and recently, roadside units that, besides many other services, register the vehicles passing by.

Probe Vehicle Data (PVD), also called floating car data (FCD), is data collected by the vehicles themselves. According to ISO 22837, probe data is designed as "vehicle sensor information, formatted as probe data elements and/or probe messages, that are processed, formatted, and transmitted to a land-based center for processing to create a good understanding of the driving environment". Since the publication of the standard in 2008, the tracking possibilities using mobile and nomadic devices multiplicated, requiring a broader definition: a device in a vehicle that is able to determine at least the vehicle GPS position, which communicates with some service application such as tolling, application servers in the Internet, roadside unit, or UMTS base stations, can in fact deliver (intentionally or not) probe data.

Probe vehicle data collection shall, however, be distinguished from conventional methods that use detectors along the roadside such as: pneumatic road tubes, piesoelectric sensors, induction loops, infrared detectors, microwave radars, closed circuit TV (CCTV) video cameras, etc.

A PVD system has the advantages that it does not need expensive infrastructure at the place where data is collected and it is able to collect data all the way along the route and not only transversally, as it is the case of fixed sensors. It is important to note that PVD data is formed to messages, which are sent to a processing server. They are not used for safety purposes and need not be exchanged between vehicles.

The ISO (International Organization for Standardization) is reading ITS Standardization. ISO is an international "de jure" standard body composed of representatives from various national standards organizations. Japanese experts have contributed to Technical committee 204 (responsible for ITS) for more than 10 years and they published some international standards about Probe Vehicle Systems. Furthermore, The U.S. Department of Transportation (USDOT); the Road Bureau of the Ministry of Land, Infrastructure, Transport and Tourism of Japan (MLIT) and the European Commission (EC) aim to advance the public sector in deploying cooperative systems and in capturing, managing, and using probe data in the management of transportation systems. The EC joined the effort in 2014, and the three regions are building on work completed to date.

On the other hand, one of the biggest concern on probe vehicle system configuration is privacy. In the Probe Vehicle System, the probe data surely contains "location" and "time" of transmitted vehicles. In order to protect the privacy, a vehicle should not be identified by the data collected. However, an authentication of the data sender is necessary to protect a probe vehicle system from a menace. The vehicle has a close relation to the data subject and the excursion of the vehicle shows an activity history of the owner. Furthermore, there is a possibility of identifying a particular vehicle on the basis of the characteristic of probe data and where it is collected. Identifying a vehicle means the possibility of disclosure of the data subject's personal information and privacy.

In addition, the phenomenal growth in smarter end-user devices and machine-to-machine (M2M) connections is a clear indicator of the growth of the Internet of Things (IoT) or the Internet of Everything (IoE), which is bringing together people, processes, data and things to make networked connections more relevant and valuable.

According to a Cisco Visual Networking Index (VNI) [4], by the end of 2014, the number of mobile-connected devices will exceed the number of people on earth, and by 2018 there will be nearly 1.4 mobile devices per capita. There will be over 10 billion mobile-connected devices by 2018, including machine-to-machine (M2M) modules exceeding the world's population at that time (7.6 billion). An important factor contributing to the growing adoption of IoT is the emergence of wearable devices. Wearable devices, as the name suggests, are devices that can be worn on a person, which have the capability to connect and communicate to the network. On the other hand, a vehicle is equipped with more than 300 sensors and moves with human beings. From the point of view of IoT, a vehicle is a wearable device.

Meanwhile, personal data and privacy information might be handled in many different ways in IoT. For example, an IoT service does not include any personal data within the data, but uses personal data to authenticate the data source when collecting the data. In this case, even if their personal data is not contained in the collected data, the user (data subject) cannot furnish data with complete peace of mind unless there are any privacy protection methods. In order to handle human activity information on the IoT, it is necessary to create basic principles and methods for privacy protection.

The privacy protection concerning a probe vehicle system is a globally important issue. For personal protection, a probe vehicle system needs to create and observe a common rule. It is the basic rules to be observed by service providers who handle privacy in probe vehicle system. This rule is aimed at protecting privacy of vehicle data senders, owners and drivers of vehicles fitted with in-vehicle probe systems, as well as their intrinsic rights and interests.

ISO/TC204/WG16 published international standard about personal data protection in probe vehicle system as "ISO 24100 Intelligent transport systems – basic principles for personal data protection in probe vehicle information services" in 2009 [5]. ISO 24100 stipulates that even if data cannot identify an individual directly, if it can do so indirectly, it should be regarded as personal data to be specified in this standard as a target of protection. It is mentioned in the OECD guidelines for personal data protection. In order to protect the privacy, a vehicle should not be identified by the collected data. However, an authentication of the data subject is necessary to protect a probe vehicle system from a menace.

2. Deployments of Probe Vehicle System

Probe information systems are developed and deployed in many countries in the world. The probe information systems were classified by two criteria such as operators and source of probe information. The operators of the system are classified by private sector and public sector. And the sources of the probe information are classified by vehicle or smart phone as shown in Table 1.

The examples of vehicle probe system in private sector are listed in Section 2.1.1. Then only one example of vehicle probe system in public sector is detailed in Section 2.1.2. In Section 2.1.3, the examples of smartphone probe deployment are presented. Beside business deployment of the probe systems, there are R&D testbed and project by Government and EU. These examples are detailed in Section 2.4.

Table 1: Business deployment of probe systems

	Probe	
	Vehicle⊠Probe	**Smartphone Probe**
Private Sector	Honda Internavi, G-book Carwings, Smart loop (Section 2.1.1)	Nokia HERE, Google Maps, TomTom HD live (Section 2.1.3)
Public Sector	ITS Spot (Section 2.1.2)	

2.1 Vehicle Probe

2.1.1 Probe System in Private Sector

Honda Internavi Premium Club

The Honda Motor Company serves Internavi Premium Club that utilizes vehicle probe (Honda calls it the floating car) since September 2003. The club members in the service have grown constantly to exceed 1.48 million subscribers as of February 2012. The Internavi floating cars of the membership collects 356,000 km cumulative distance. Internavi Route provides the route navigation system to the users, based on their preference using VICS information, current information from the Internavi floating car, and road section information recorded in the past data.

The probe data is uploaded from Honda's navigation system of the club members to the server in the form of the time required to pass a road section (road link). The uploaded data is used when the users request the route navigation and real-time traffic information. Initially, Honda's navigation system was only a corresponding device; however, Smart Loop (See detail below) by Pioneer Corporation was granted access for "Floating Car Data" to use the probe data mutually in August 2008. The main communication media is the cellular systems. The "Linkup free" service was launched in 2010 and the users do not need to pay the cost for the communication under certain condition using Honda's navigation system. The service can be realized thanks to the deployment of mobile virtual network operator (MVNO). Honda is the MVNO of the "Linkup free" service using the cellular network of Softbank Mobile.

Carwings

The Nissan Motor Company starts the service of Carwings by expanding the previous service of "CompassLink" in 2004. Carwings provides the traffic information dissemination service from the vehicle probes since November 2006. The service provides the traffic information of the zone that is not covered by VICS thanks to the information periodically sent from the members' vehicle.

Real time traffic is estimated with the probe vehicle information in addition to VICS information. Also, Carwings predicts the traffic condition and traffic jam in the future about two hours by comparing the estimated real time traffic, the VICS information and the past pattern of traffic jam. Initially the Nissan navigation system are only corresponding devices; however, Smart loop by Pioneer Corporation exchanges the probe information with Carwings for mutual benefit since 2009.

G-BOOK mX / T-connect

The Toyota Motor Corporation starts the "G-BOOK mX" service consisting of traffic information system using vehicle probe and map update system since May 2007. "G-BOOK mX" collects the driving data with high frequency by the dedicated communication device called data communication module (DCM). Like Honda "Internavi Linkup free", Toyota becomes a mobile operator as MVNO of DCM using the KDDI's cellular network. Toyota claims that the system can collect sixty times more driving data (from Toyota's previous service). The probe vehicle data taken at every 10 meter is uploaded to the G-BOOK center and the information is provided to the users along with VICS's current information and statistics information.

The iPhone and Android smart phone application was released in 2010 in order to expand the telematics service from the Toyota's navigation system to the smartphone. Toyota also announced "T-Connect" in June 2014 along with G-BOOK. In T-Connect, the Toyota navigation systems and the smartphone can connect "Toyota smart center" in the cloud system in the Internet in order to use the application and extension.

The application is distributed in the application store and the user can install the application from the store. The system is an open environment (TOVA, Toyota Open Vehicle Architecture) and the software development kit (SDK) is provided to the application developers.

Smart Loop

Pioneer Corporation announced "Smart Loop" through which the users obtain the valuable information by exchanging their own knowledge with each other among the drivers and makers. In order to enable the smart loop, Pioneer releases the navigation system with the record type probe function. The information dealt in "Smart loop" includes the users' knowledge and experience about driving. Therefore, the exchange information is not limited to the driving data. For example, at which road sign the vehicle is to stop in a certain route, which parking the driver uses for the destination, stop point, registered place data, fuel efficiency management data and voice recognition data. Pioneer provides the incentive to the informant with various services such as drive mile service and the driver's report.

Initially, the service started as record type probe system, that the data is recorded internally to the dedicated hardware or USB memory during the driving and then the user can upload the data to the server via PC. Since 2007, the real time probe system is released using cellular communication media and dedicated communication module. This system supports to upload maximum driving recodes of 50 km at once. At the same time, the system estimates the traffic condition in the area that is not covered by VICS from the real-time probe system using 1-hour old record and the past 90 days' history.

Smart loop becomes one of the most advanced cooperation service with the other probe services, which is interoperable with Honda

Internavi since 2008 and with Nissan Carwings since 2009. Pioneer and NTT DoCoMo jointly released Android application named "DoCoMo drive net powered by Carrozzeria" as well as the dedicated cradle in 2011. "Smart loop eye" was introduced in 2013, sharing the photos taken by the probe vehicle in order to check the traffic condition such as traffic jam.

2.1.2 Probe System in Public Sector and Testbed

ITS Spot

An example of probe system deployment in public sector is utilization of probe vehicle data with ITS Spot installed by Japan Ministry of Land, Infrastructure, Transport and Tourism (MILT). ITS Spot consists of short range dedicated communication infrastructure installed in about 1600 points in the main roads in Japan (see detailed points in Fig. 6) and ITS Spot compliant on-board unit from the private vendors including the navigation system. ITS Spot supports dynamic route guidance, safe driving support, Electronic Toll Collection (ETC) System and Internet services.

Today, Japan MILT puts in an effort to realize the improvement of efficient road management by collecting the probe data recorded in the car navigation systems from different vendors. ITS Spot compatible device records the position, time and speed in every 200 meters or when the heading of the vehicle changes over 45 degrees. In addition, it also records position, time, speed, acceleration and yaw rotation angle when it detects the acceleration over certain degree. The recorded data is stored in the hardware of the on board unit at the time and uploaded to the center when the vehicle passes the ITS Spot infrastructure.

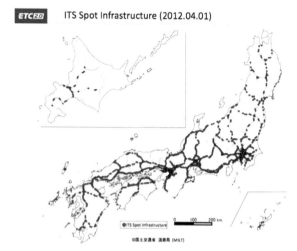

Figure 6: ITS Spot deployment (from the MILT website).
http://www.mlit.go.jp/road/ITS/j-html/spot_dsrc/tenkai.html

2.1.3 Smartphone and Other Probe

The movement of a cellular phone can be detected in the cellular base station where the phone connects. Road traffic information can be generated with the data in the base station and the other information such as road network information. These systems were investigated and developed in 2000s. California University, Nokia and NAVTEQ (bought by Nokia in 2007) started the Millennium project for the smartphone based probe information system. Thus the number of smart phone probe grows exponentially after the release of the smartphone probe application by several traffic information providers in US. The user of the probe application can receive the traffic information for free.

Nokia HERE

Nokia HERE is famous example of smart phone probe system, that provides real-time traffic information service. Nokia HERE is based on the service provided by NAVTEQ, that was the biggest map company in US. NAVTEQ launched the first commercial service of car navigation and real time traffic information in 2004. After joining Nokia, it develops the world wide service based on the knowledge and know-how acquired in more than 25 years of map service and 10 years of traffic information service.

Today, HERE Traffic is a real time traffic information service from the vehicle probes and has served in 41 countries and regions. HERE Traffic Patters provides travel time, average speed and accident points from recorded probe data in 81 countries and regions. Amazon and Yahoo are the clients of HERE service. The HERE platform processed more than one hundred million traffic information requests per day. HERE Traffic collects more than 20 billion vehicle probe data per month and 60% of the probe data from vehicle is less than two minutes' freshness. The contents of the real time traffic information is updated within 1 to 3 minutes depending on the service.

The information is provided by Radio Data System/Traffic Message Channel (RDS-TMC), digital radio as well as HTTP (Web). The accuracy of the information is verified 150 times in 180 thousand km driving distance, and 6000 hours of actual driving test per year. These tests are analyzed and used for the improvement of the service.

TomTom HD Traffic

TomTom HD Traffic is one of the most known services in Europe. This service collects the driving history from Portable Navigation Device (PND) and phone and generates the traffic information, based on "MapShare" by Tele Atlas (has been a wholly owned by TomTom since 2008). TomTom HD Traffic provides the surrounding information of the vehicle every two minutes, detects traffic jam caused by heavy traffic and proposes better alternative routes to the users.

Google Maps

Google traffic is the most used traffic information service from smart phone users. The user can access traffic information via web browser for free. The service was launched in some cities and regions in 2008 and has expanded to many countries today. Google traffic shows the five levels of traffic condition to each road in the service. The level is determined by the algorithm in Google based on the location and speed information from the Android smart phone where the user activates the "my location" service.

2.2 R&D Testbed and Project by Government and EU

Beside the business deployment, Government and EU encourage the research and development of more advance probe services. The USDOT initiated the connected vehicle research testbed in Michigan, Virginia at the Turner-Fairbank Highway Research Centers (TFHRC), California, Florida, New York, Minnesota and Maricopa County.

In Europe, many European and national projects are funded for the field operational tests (FOTs) for the probe information system, DRIVE C2X, In-Time, ROADIDEA, Track&Trade, Viajeo, PRESERVE, FREILOT, INTRO Intelligent Roads, MOBINET, SimTD, and ITS Corridor.

Table 2 summarizes these R&D testbeds and projects by Government and EU. Please refer [3, 6] for more detail for these testbeds and projects.

Table 2: R&D testbed of Government and EU

Testbeds in United States	Projects in European Union and European countries
The Connected Vehicle Testbeds in Michigan, Virginia at the Turner-Fairbank Highway Research Centers (TFHRC), California, Florida, New York, Minnesota and Maricopa County (See more detail in [3])	FOTsis, DRIVE C2X, In-Time, ROADIDEA, Track&Trade, Viajeo, PRESERVE, FREILOT, INTRO Intelligent Roads, MOBINET, SimTD, and ITS Corridor (See more detail in [6])

3. Standardization on Probe Information Systems

3.1 Overview of Standardization Activity of Probe Information Systems

Commercial service using vehicle probe information has started in 2004 by Japanese auto manufacturer, Honda Motor Co., Ltd. Since then, various services have been developed.

Standardization in ISO started in early 2000s. In 2009, the first ISO standard ISO 22837 Probe data was published. Special care was taken so that the standard does not contradict existing probe vehicle information systems. Standards concerning probe data transmission timing (Probe data reporting management ISO/TS 25114:2010, Event based probe data ISO/TS 29284:2012) were published. Recently, standardization activity on service architecture are in progress. In addition to ISO standardization, SAE (Society of Automotive Engineers) defines Dedicated Short Range Communications (DSRC) Message Set Dictionary J2735. Probe data is defined in J2735 as one of messages.

Since the probe vehicle information system has some form of identification, privacy protection is an important issue. ISO 24100:2010 Basic principle of privacy protection in probe vehicle information services was published in 2010. This standard is based on the OECD guideline of transborder data transfer 1980. Maintaining the standard of privacy protection criteria on probe vehicle information systems is in progress.

3.2 Published Standards

In this subsection, published standards are described in three categories: data, management and privacy protection. An overview of probe vehicle information system is given in Fig. 7. ISO standards currently available and also work in progress is also drawn with the image of scope.

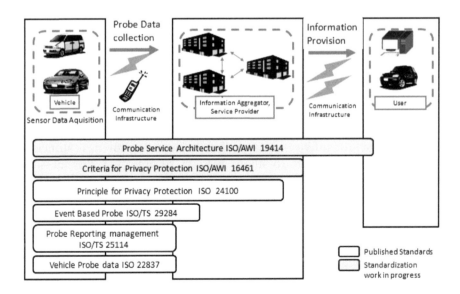

Figure 7: Overview of probe vehicle information systems.

3.2.1 Data Related Standard

Two standards have been published in the field of probe data definition. One is ISO 22837:2009, which define probe data and context architecture, another is SAE J2735, which define messages for Dedicated Short Range Communication (DSRC).

3.2.1.1 Probe Data (ISO 22837:2009)

ISO 22837:2009 Vehicle probe data for wide area communications was published in 2009. According to ISO, in this standard, reference architecture for probe vehicle systems and probe data are defined. This architecture provides a general structure for probe vehicle systems using various communication mediums. The information model which provides the logical structure of entities and concepts involved in probe data is also defined. Core data element, which is required in all probe messages, is defined. This element consists of location and time when location is sensed. Some examples of probe messages are also described.

3.2.1.2 Data Dictionary (SAE J2735)

SAE (Society of Automotive Engineers) published J2735 Dedicated Short Range Communications (DSRC) Message Set Dictionary in 2006. In this document, probe vehicle data is defined as one of messages for DSRC. Since its first publication, it was revised twice, and the current version was published in April 2015. According to SAE, the SAE J2735 standard comprises a complete list of all dialogs (message exchanges), messages, data frames (complex elements), and data elements (atomic elements) which are used in the message set. This standard specifies a message set, and its data frames and data elements specifically for use by applications intended to utilize the 5.9 GHz Dedicated Short Range Communications. This standard, however, has been designed, to the extent possible, also to be of potential use for applications that may be deployed in conjunction with other wireless communication technologies. This standard therefore specifies the definitive message structure and provides sufficient background information to allow readers to interpret the message definitions properly from the point of view of an application developer implementing the messages according to the DSRC standards.

3.2.1.3 Vehicle Data Interface (World Wide Web Consortium)

Data interface for acquiring data from vehicle systems through the web browser is being developed in World Wide Web Consortium (W3C). A Draft, Vehicle Data Interfaces, has been created by Automotive and web platform business group in December 2014. Vehicle Data Interfaces and Vehicle Information API were created, based on the achievements by Genivi, Tizen, Webinos project and QNX. An automotive working group for creating recommendations, based on Vehicle information API and Vehicle Data Interface, has been established.

3.2.2 Data Management

In probe vehicle information systems, there are two types of modes in transmitting probe messages from vehicles. One is the message in response to the request (polling) from the probe message processing center. Another is the message voluntarily sent from the vehicle. This mode comprises two types, periodically sending messages and a message sent when an event occurs. In the system of periodic message sending, management, such as changing period, start/stop sending message, is needed. With this background, two ISO TS (Technical specification) have been published.

3.2.2.1 Probe Data Reporting Management (ISO/TS 25114:2010)

ISO/TS 25114:2010 Intelligent transport systems – Probe data reporting management (PDRM) has been published in 2010. According to ISO, this technical specification provides a common framework for defining probe data reporting management (PDRM) messages to facilitate the specification and design of probe vehicle systems and gives concrete definitions of PDRM messages. This technical specification also specifies reference architecture for probe vehicle systems and probe data which incorporates PDRM, based on the reference architecture for ISO 22837, and basic data framework for PDRM instructions, which defines specifically necessary conditions for PDRM instructions and notations of these instructions (in XML).

3.2.2.2 Event Based Probe Data (ISO/TS 29284:2012)

ISO/TS 29284:2012 has been published in 2012. This technical specification defines various aspects on event based probe data. Instead of sending probe messages in a fixed time or distance interval, this technical specification deals with the system that sends a message when a pre-determined event, such as hard breaking, is detected.

This technical specification specifies basic data framework of event based probe data reporting as well as reference architecture. It also defines an initial set of event-based probe data elements, which are used typically in application domains such as traffic, weather and safety. It serves as a supplement to ISO 22837:2009 and specifies additional normative data (probe data elements) that are delivered by an event-based probe data system.

3.2.3 Privacy Protection

The probe vehicle information system gathers location and time and other information from a vehicle. A typical probe vehicle information system does not need to distinguish individual data points. Traffic information, for example, creates the traffic condition of a road section from data gathered as probe data. Special care has to be taken in handling probe data/messages in order not to violate vehicle owner's privacy right. A standard has been developed for this purpose.

3.2.3.1 Basic Principle for Privacy Protection (ISO 24100:2010)

ISO 24100:2010 Intelligent transport systems – Basic principles for personal data protection in probe vehicle information services has been published in 2010. This standard is aiming at setting basic rules for vehicle probe information system providers. This international standard is aimed at protecting the personal data as well as the intrinsic rights and interests of probe data senders, i.e., owners and drivers of vehicles fitted with in-vehicle probe systems.

ISO 24100:2010 states the basic rules to be observed by service providers who handle personal data in probe vehicle information services. The base of this standard is the OECD Guidelines on the Protection of Privacy and Transborder Flows of Personal Data, 1980 and the data protection legislations of individual country.

3.2.3.2 Criteria for Privacy and Integrity Protection (ISO/AWI 16461)

ISO/AWI 16461 Intelligent transport systems – Criteria for privacy and integrity protection in probe vehicle information systems is in a development stage. (AWI: Approved Work Item) Based on ISO 24100:2010, this work item aims at providing criteria for privacy protection in probe vehicle information systems. This work item has been approved as new work item in 2014.

3.2.4 Probe Service Standards

Probe data is widely used in the market. There is an activity of categorizing these services.

3.2.4.1 Service Architecture (ISO/AWI 19414)

ISO/AWI 19414 Intelligent transport systems – Service architecture of probe vehicle systems is a work item in progress. This was approved as work item in September 2014. This aims at classifying the ITS services which uses probe vehicle information. The service framework of probe vehicle information systems will be established. In addition, definitions of service domains that utilize probe data from probe vehicle systems will be given.

References

1. Report of feasibility study about social impact of ITS. 2002. Japan Automobile Research Institute (JARI).

2. International Standardization Organization. 2009. Vehicle probe data for wide area communications (ISO 22837:2009).
3. US-Japan Collaborative Research on Probe Data: Assessment Report. 2013. Final Report. Publication Number: FHWA-JPO-13-091.
4. Cisco Visual Networking Index: Global Mobile Data Traffic Forecast Update 2014–2019 White Paper, http://www.cisco.com/c/en/us/solutions/collateral/service-provider/visual-networking-index-vni/white_paper_c11-520862.html
5. International Standardization Organization. 2010. Intelligent transport systems – Basic principles for personal data protection in probe vehicle information services (ISO 24100:2010).
6. Sandford Bessler, Thomas Paulin. September 2013. Literature study on the State of the Art of Probe Data Systems in Europe. FTW Telecommunications Research Center, Vienna. http://fot-net.eu/literature-study-on-the-state-of-the-art-of-probe-data-systems-in-europe/ Last access on June 27th, 2016.

Complex Infrastructures: The Benefit of ITS Services in Seaports

Paolo Pagano[1]*, Mariano Falcitelli[1], Silvia Ferrini[2], Francesco Papucci[2], Francescalberto De Bari[2] and Antonella Querci[2]

[1]Joint Laboratory of Advanced Sensing Networks & Communication in Seaports,
Consorzio Nazionale Interuniversitario per le Telecomunicazioni (CNIT),
Livorno, Italy
[2]Development and Innovation Department, Autorità Portuale di Livorno
Livorno, Italy

Abstract

Seaports are genuine intermodal points of interest connecting seaways to inland transport facilities such as roads and railways; a seamless and structured way of interconnecting modes of transport is beneficial to both freight and passengers.

In the former domain, seaports are intermediate nodes in the logistic chain: the availability of services pointing to a prompt and effective handling of goods is a good performance indicator to attract logistic operators and turn to be competitive in the market. Along the same line, in the latter, informational services can support passengers on the move (helping people with disabilities or promoting touristic offers).

Vehicular communications will play a more and more significant role helping passengers to commute as in Ro-Ro terminals, turning to be an enabling technology for a set of diversified vertical implementations. The real challenge is therefore that of considering Intelligent Transport Systems (ITS) as part of the ICT innovation agenda, integrating ITS services in the control premises of seaports and letting ITS novel solutions contribute to improve the performance indicators of seaports and mitigate the risk level in relevant areas.

*Corresponding author: paolo.pagano@cnit.it

In this chapter the case study of the Livorno port will be considered, the design and implementation plan of novel ICT services for a variety of final users from the private sector as well as from the public (governmental, authority) sector.

1. Directives and Regulatory Constraints in Implementing ITS in Seaports

Seaports transfer passengers and goods coming from the ships to further transit points and final destinations complying with international maritime regulations. These regulations affect the procedures and the (digital) administrative transactions allowing a certain person or piece of goods to reach the next point of transit.

The main directives in the European Union regulate the information set related to the (arrival/departure) schedule time of the ships (Directive 2002/59/EC), customs controls for travelers (Regulation 2006/562 EC), notifications of dangerous freight (Directive 2002/59/EC), notification of waste and residuals (Directive 2000/59/EC), notification of security to entry ports (Regulation 725/2004 EC) and entry summary declaration for customs clearance (Regulation 450/2008 EC). These regulations sometimes implement more general provisions for the International Maritime Organizations members practiced worldwide.

All directives tend to implement electronic exchange of information, avoid duplications, force member states (EU and IMO) to set up a Single Point Of Contact (SPOC) accessed through a National Single Window (NSW) instead of keeping alive dozens of legacy information systems run by independent regulatory and control agencies (see Fig. 1).

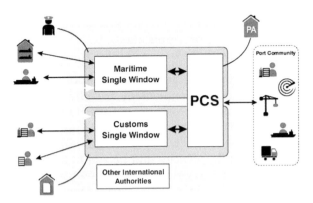

Figure 1: Concept of National Single Windows (maritime and customs) handling port operations, customs procedures, technical controls and certification procedures. (Courtesy of Actual I.T. d.d., Koper (SLO), http://www.actual-it.eu/ Last access on May 12th, 2016.)

Each port is supposed to set up an information system, known as Port Community System (PCS) to connect to the SPOC established at national level; private and public institutions use PCS to enable intelligent and secure exchange of information, to improve the efficiency and competitive position of the seaports, to automate and smooth port logistics processes through a single submission of data and by connecting transport and logistics chains. At the EU scope, NSWs are networked (see the SAFESEANET platform [1]) in order to exchange information from member state to member state, fulfilling the target of implementing corridor-wide information systems. Along this line, the Directive 2010/65/EU supports the so-called "ONCE paradigm", i.e. the approach of implementing a single electronic transmission of data and a single control declaration formalities for ships arriving and for ship departing/arriving in/from ports of member states.

Seaports are, therefore, domains where the international community fosters innovative electronic data exchange systems leading to paperless administration, harmonization, integration and simplification of processes. How ICT solutions for intelligent transportation systems can intercept the case studies of seaports is opening an interesting debate at the EU level.

The Directive 2010/40/EU fosters the implementation of ITS solutions to increase the efficiency of the transport sector at system level, i.e. to provide proper optimized mobility management along the whole (intermodal) chain. The Commission points at providing Europe with a coherent, inter-modal transport system, finally solving the uncoordinated deployments actually missing 'geographical continuity of ITS services throughout the Union and at its external borders having set the following priorities:

- Optimal use of road, traffic and travel data;
- Continuity of traffic and freight management ITS services;
- ITS road safety and security applications;
- Linking the vehicle with the transport infrastructure.

Seaports will contribute in 'achieving a European transport system that is resource-efficient, environmentally friendly, safe and 'seamless' for the benefit of citizens, the economy and society' given their genuine nature of intermodal points at the edges of trans European corridors.

As applicable to the whole transport sector, infrastructures are key elements to transfer goods, vehicles and people from a specific point of distribution to another one, in a short time and with a high level of safety and security. In particular, in ports featuring container terminals the presence of efficient facilities allows to receive large vessels and handle different types of cargo, reducing the transit time. The same provisions can be easily ported to roll-on roll-off terminals and points of passengers transportation. However, in the internet era where innovation plays an important role in the competition perspective, the presence of appropriate facilities is a necessary but not sufficient condition.

Seaports also generate a huge amount of heterogeneous data that must be gathered, transferred and processed by the port communities.

2. Seaports in the ITS: The Use Cases

A seaport exploiting a sensing and communication layer pervasively will enable it to set up applications relevant for the logistics sector. It will be beneficial to interconnect the seaport with the other points of interest in the logistic and transportation chain especially considering the intermodal perspective.

From a managerial and financial perspective, a seaport is ranked considering a set of Key Performance Indicators (KPI) coming from the operational, environmental, safety and security aspects of the Service Level Agreement (SLA):

- Transit time is one of them particularly relevant for ports gateway which are supposed to handle mainly freight and passengers in transit;
- Pollution levels are also considered for the health and wellness of people working or living nearby a port; this is especially relevant for historic ports that are usually located in proximity of the urban settlement;
- Security of managed information is considered important especially when information flows from system to system under the responsibility of different entities;
- The safety of the workers and visitors should be analyzed in real-time (i.e. risk level assessment) along the processes involving specific people, vehicles and zones.

"Users" (i.e. companies and/or passengers) will choose a certain seaport not only for the availability of interesting (multi and inter-modal) commercial links, but also for the effectiveness of the logistic offer, with particular attention to time and costs handling.

On the other hand, the "Regulators" will support larger volumes provided no more pollution, no increased risks, no breach for frauds occur.

Both of them will benefit from an integrated ICT system capable of delivering a certain function effectively, enforcing regulations and keeping track of the use of resources and reducing level of pollution and risks.

The creation of automatic procedures is the main challenge of the Ports Gateway in order to both reduce transit time of freight and passengers and offer dedicated services to specific users. In this perspective, monitoring drivers and vehicles enables them to provide automatic authentication procedures in order to offer both essential and value-added services only to authorized users (i.e. access to port and terminal gates, avoiding congestion and getting rid of starvation-prone procedures). In multi-purpose ports a lot of means (e.g. ordinary cars, heavy trucks and dock vehicles) roam inside the port area, moving persons and (dangerous goods) from/to large vessels at berth.

- All drivers and vehicle information inside the port area, concerning authentication, position and other relevant events can be recorded and processed in real-time. Starting from a driver/vehicle authentication, specific paths and services are suggested. The driver (hauler or visitor) is assisted in the navigation, towards the freight loading/unloading bay or the destination berth in accordance with the user profile target. During the path the driver is notified about any impediments or risks he will face (to accomplish the authorized task) and workers or other entities (vehicles, points of interest) are notified of the vehicle maneuvers.
- The pollution level must be continuously monitored in industrial and urban areas following the regulations on sustainable growth in force by national and European bodies. In this perspective, environmental sensors deployed in the port area will monitor environmental conditions and allow the said bodies to take action when health thresholds are overcome.

In addition to vehicles, in a multi-purpose port, developing cargo monitoring systems can also be beneficial for the seaport competitiveness.

- Public bodies can also exploit such systems intervening in case of emergencies and to detect cargo integrity and accountability problems.
- The container sensing history (i.e. temperature, acceleration, light exposure, etc.) can be downloaded for off-line quality checks. Cargo conditions can be detected by On-Board Units (OBU) installed on the container/semi-trailer and connected in real-time with the infrastructure via Road Side Units (RSU). Eventual alarms or specific notices can be recorded in the seaport data repositories and eventually displayed in monitoring systems and video walls.

Besides industrial processes and activities, seaports are ecosystems where visitors and workers live. Thus port authorities and haulers must adopt suited solutions for safety reasons. Risk level depends on concurrent conditions (e.g. type of cargo handled, type of equipment used, etc.) and real-time monitoring of the risk level is needed.

- Exploiting such service, regulators and emergency operators (i.e. policemen and the fire brigade) can promptly intervene in specific areas in case of need.

- It is also recommended to prevent these types of emergencies, having previously assessed some specific risks.
- Port workers and users have to be constantly informed about their possible risks. When a lot of variables are pervasively gathered, managed and processed, risk-level can be calculated with a high accuracy (e.g. induced by some dangerous goods or some detected failure on cranes). To this extent OBUs send real-time information to RSUs that relay to the infrastructure and to specific supervisors in shift. All the information has to be elaborated and alarms distributed in real-time, to allow for a prompt intervention.

3. Design of the ICT Info-structure and Services: A Cloud of Ports

A harmonized info-structure resulting from proper integration will host the software services needed for operational, commercial, and planning activities. Hosted into this info-structure useful information will be aggregated into knowledge for supporting decisions and to enable processes.

As ports are distributed, multi-user information systems, they feature an Infrastructure (IaaS), a Platform (PaaS), and a Software (SaaS) layers as in popular state-of-the-art clouds.

Hopefully a set of new applications based on emerging ICT technologies will be integrated at the SaaS layer of the cloud.

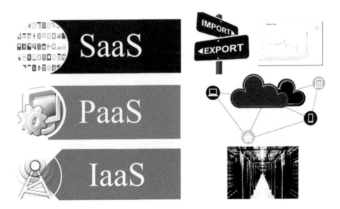

Figure 2: A cloud-shaped layered information system.

To properly set up the Infrastructure as a Service (including Network as a Service) layer, the following recommendations must be addressed:

- Connect the port in itself, i.e. setting up a LAN of the communities that are based on the port landside;
- Connect the port with other nodes along the transport and logistics chains, i.e. setting up a corridor-wide internet of the (member states) communities;
- Pervasively monitor and control port activities by gathering data from humans and machines (e.g. sensors, vehicles) interconnected through the LAN or the internet.

At the PaaS layer, a middleware of services will offer the custodial access, and retrieval of information coming from the port, the internet of port communities, and external repositories.

All applications, notably those related to PCS, are intended for the SaaS layer, relying on the information coming from the middleware, eventually retrieved from the internet of port communities, and external repositories.

4. The Port of Livorno

The Livorno port (see Fig. 3), favorably situated in western Tuscany, plays a major role in the European internal trade, ensuring frequent and fast maritime connections to southern European countries, as well as in the EU external trade, thanks to its well-established linkages to northern African countries and the Americas.

Figure 3: Aerial view of the Port of Livorno.

Livorno is also considered a pivotal node in the logistic chains linking the Mediterranean to central-east Europe.

The port of Livorno is an ideal reference for implementing ICT solutions oriented to sustainable growth because:

- it is a mid-size port with a complex infrastructure yet expected to be relatively simple to interact with;

- the port and city are intertwined, as it often occurs in ancient (16th century) ports;
- it is multi-purpose, including a variety of operations – containers, cargo via Ro-Ro, ferries and cruises.

The Livorno Port Authority (APL) is a regulatory rather than operational organization, hence able to consider and participate in development of novel ICT technologies, also exploitable in the mid- to long-term.

Livorno ranks first in Italy for new car traffic (second in the Mediterranean Sea) and first for pulp and Ro-Ro traffic. Livorno, along with the Guasticce Freight Village nearby, is listed as core node of the Trans-European Network of Transport (TEN-T) and is part of the core network "Corridor I", which links the Mediterranean Sea to the Scandinavian peninsula.

Livorno has direct access to the highway to Florence, a public infrastructure owned by the Regional Government and operated by a consortium of private industries.

The Livorno settlement is shown in Fig. 4 together with its main connections to highways, railway resources (i.e. lines and stations), and the off-site buffer offered by the Freight Village.

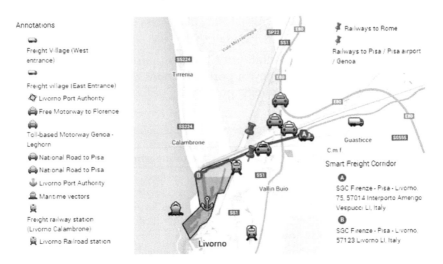

Figure 4: The intermodal facilities for freight and passengers in the Livorno area. (See an annotated map at: https://goo.gl/O6l2SZ. Last access on May 12th, 2016.)

4.1 The Port Master Plan

The port layout is currently undergoing important changes, due to the approval of a new port master plan which foresees a deep organization of port land and planning in coming years. The most relevant improvement is represented by the new "Europe Dock" (in Italian "Darsena Europa"), a large infrastructure made up by two large terminals, one dedicated to the newest container vessels, with a capacity up to 18,000 TEUs and the other specialized in the Ro-Ro and passenger traffic (see Fig. 5). To the best of the authors' knowledge, this is the largest infrastructure investment of the last decade in Italy (approved project is worth of about 800 million Euros from public and private investment).

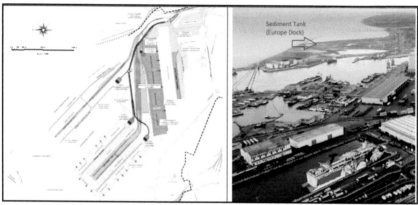

Figure 5: Project plans and work in progress in Livorno for the new "Europe Dock".

Through the implementation of the new port master plan, greater achievements in the efficiency of port operations are expected as well as the strengthening of the strategic positioning of the Livorno logistic node in the Mediterranean.

Table 1: Figures of the Port of Livorno

Current records of the port	Major investment (Europe Dock)
• First Italian port for Ro-Ro traffic • First Italian port for new cars traffic (2nd in the Mediterranean) and for pulp • Major Italian port for general cargo • Connected with national and European road and rail network • Multipurpose port with broad variety of cargo, unitized, bulk and general cargo	• Total surface: 105 ha, entirely through land reclamation • Container terminal surface: 55 ha • Logistic facilities surface (railway station, marshalling yard, roads, services areas): 38 ha • Rail track suitable for 750 m module trains • Berths: 20 m depth, up to 900 m long • Turning basin: 600 m diameter, 20 m depth

As of January 2016 a new electrified railway connection has enabled the direct forwarding of block trains from the already existing container terminal (in Italian "Darsena Toscana") to the rail line Genoa-Rome, thus avoiding shunting operations and lowering the transport costs (see Fig. 6).

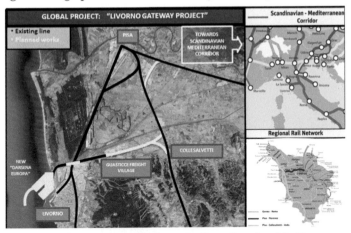

Figure 6: Railway connection from the Scandinavian-Mediterranean corridor to the Port of Livorno.

4.2 The Road Network and the Dry Port

The Port of Livorno is directly connected with the Florence – Pisa – Livorno highway (locally known as FiPiLi) at the mouth of the "Arnaccio" channel in the Tyrrhenian Sea.

The highway connects Livorno to Florence and Pisa having a total extension of 100 kilometers of dual carriageway with central barrier. The highway is integrated with three important national motorways:

- Pisa – A12 (Livorno/Genoa) by a link road at the highway edge;
- Firenze Scandicci – A1 (Milan/Naples) via a junction;
- Collesalvetti – A12 (Livorno/Genoa) via a junction.

It is east-west oriented widely recognized as an asset of the Tuscany region that delegates its management to local county administrations. The highway is also serving the airport of Pisa, the main Tuscan hub featuring a yearly traffic of more than four million passengers. With 4.6 million passengers per year, Pisa is ranked 10th in the list of busiest airports in Italy (82nd in Europe).

The dry port is an independent infrastructure (managed by a Public Private Partnership), located 5 km eastwards of the port, accessible directly via the highway. It is a fully cabled environment consisting of warehouses (more than 300,000 sqm), yards (about 700,000 sqm) also including a railway terminal (more than 125,000 sqm) directly connected to the port docks.

In the port masterplan the ensemble port – dry port are to be considered as a unique entity capable of delivering services to the industries involved in logistics activities seamlessly.

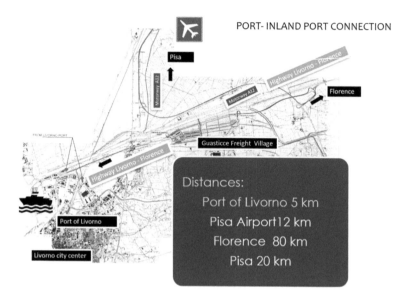

Figure 7: The Port of Livorno and the dry Port of Guasticce connected via railways and motorways.

4.3 The Port Innovation Roadmap

On 21st of January 2013, APL and the Italian Ministry of Transport have signed a Memorandum of Understanding to consider the Port of Livorno as host of piloting experiences in the domains of Innovation, Energy, and Training. To this extent APL has got a certain number of innovation projects as a promising background [2-7].

The Port Authority has followed up this opportunity to foster innovation and has signed a framework agreement with the Italian National Inter-university Consortium for Telecommunications (CNIT) to implement proper ICT technology transfer actions and let the port system be:

- greener, safer, and more efficient in touristic and industrial activities;
- more (inter-modally) connected as it is expected from its "gateway" ambition.

To implement the ICT agenda CNIT and the Port Authority decided to set up permanent lab facilities, dedicated to the technology transfer and technical validation in the field on Oct. 14th, 2015. The "Joint Laboratory of Advanced Sensing Networks & Communication in seaports" will:

- provide a continuous and effective presence of CNIT researchers at the port;

- provide an accessible and ready testing and experimentation ambient to Italian and European partners either industries or academia;
- incrementally deploy a Service and Control Room for the Port starting from the most innovative applications;
- guarantee an outstanding level of education and training for the (forthcoming) port operators.

5. Setting up the Ground for the Community

5.1 Connecting the Port in Itself

The main pre-requisite to develop new IT services in transport and logistics is to connect the port pervasively by means of a high bandwidth backbone. That is the case of the Livorno port, already fully cabled via optical fiber in the 2000s. Third parties (both commercial and institutional) can connect to the network owned by the Port Authority and operated by a commercial internet software provider, also in charge of managing internet connectivity and guaranteeing quality of service.

Figure 8 depicts the coverture of the fiber on the port landside shown. The physical layer is also structured in zones connected to isolated server and router dedicated rooms.

Figure 8: The Fiber Optics map on the Livorno port landside. The Authority headquarters (Scali Rosciano and Maritime Terminal) are also marked in the figure as the center of mass of the fibers branches.

5.2 Connecting the Docks with the Road and Rail Network

Industrial (Gateway) ports are populated by heavy trucks and trains in charge of routing the piece of freight to their next node in the supply chain.

Infrastructure development is requested to allow for an effective routing. It is also required to integrate the transport and logistic software portals to assist and support the hard work of unloading goods from the ships to the next carrier (or vice versa).

As in computer science, a good processor is generally featured by well-dimensioned buffer, designed to accommodate incoming and outgoing data streams, taking into account different traffic patterns.

In seaports, the buffer is provided by the docks themselves and the freight village (organized in stores and parking yards); a good design of the telecommunications infrastructure would therefore target to extend the port landside fiber optics to include the freight village.

In the case of the Livorno seaport, this buffer is localized 6 km to the east, directly reachable both via the Livorno-Florence highway (for Ro-Ro traffic in the so-called motorways of the sea) and via a dedicated railway corridor expected to turn operational by the end of 2016.

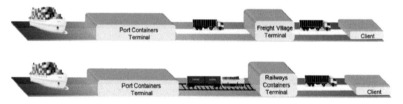

Figure 9: The port logistic chain via the motorways of the sea or via the railways.

These links establish a freight corridor eligible for smart services (and hopefully effective for customs clearance). The port-freight village ensemble must therefore be structured as an extended LAN on top of an integrated physical layer.

To this extent the regional government and the Port Authority have funded the design and the implementation of a wired broadband link along the track of the highway so that the port will encompass its "freight buffer" by the end of 2016.

6. Setting up the Ground for the Community

6.1 The Tuscan Port Community System

A port community system is a neutral and open electronic platform enabling intelligent and secure exchange of information between public and private stakeholders in order to improve the efficiency and competitive position of the seaports communities. Similar platforms are also popular for airports.

In Tuscany an open platform, the Tuscan Port Community System (TPCS) has been implemented featuring full interoperability with the maritime information system managed by the National Harbor Master Corps and the customs information system managed by the Customs Agency (also used by the Fiscal Service).

Figure 10: A photograph of the Tuscan Port Community System web portal. Visit http://tpcs.tpcs.eu. Last access on May 12th, 2016.

The system is used by carriers, terminal operators, shipping lines and control agencies to manage the work flow around the shipment along the supply chain.

In what follows, special considerations about the most innovative (and piloting) aspects of the Tuscan Port Community System are discussed:

- TPCS permits carriers to run customs clearance procedures while the freight stock is over the sea (i.e. for import) simplifying the usual paper work foreseen when the freight is unloaded from the ship (or when the Ro-Ro cargo disembarks).

  Exploiting the output of some European projects and relying on the connected infrastructure encompassing the seaport and the dry port, a pilot has been implemented and demonstrated permitting carriers to follow customs clearance procedures making use of special RFID seals, opened when the cargo finally reaches the freight village. The reverse, that of export, has also been tested, i.e. to prepare a cargo truck for boarding (with customs clearance) while it is still in the freight village settlement.

  The objective of the authority is to turn these pilot initiatives into ordinary, legally allowed good practices by fully integrating the

TPCS with the National Logistic Platform (an ICT tool designed by the Ministry of Transportation). The platform disciplines the transit of freight along the Italian territory.

It foresees the use of fully connected trucks (equipped with special On-Board Units) along "secure corridors" like that of Livorno opening the sight to innovating technologies in the domain of intelligent transport systems, for what refers to vehicular networks.

- The Customs Agency is considering Machine-to-Machine (M2M) communication in non-authorizing transactions for the work flow management of freight shipment. Acquiring the Automatic Identification Signal (AIS) from a ship approaching the harbor as its final destination, the Port Community System will automatically check the compliance with customs clearance regulations for the freight aboard. If the compliance is certified, the freight (either containers or Ro-Ro cargo) will be set to a pending status until the human operator validates and completes the transaction.

This case study suggests considering smart sensing and M2M as a general enabler of customs procedures in seaports. To this extent the exploitation of a pervasive data collection layer together with appropriate aggregation procedures (providing semantics to raw data as will be discussed in Section 6.2) will allow to feature more and more effective services in the TPCS.

6.2 MONI.CA: The Livorno Port Monitoring System

Sitting atop the PaaS layer, MONI.CA is a strategic "convergence" platform aimed at both processing and visual rendering of data coming from the sea (through AIS signaling) and the port landside (see Fig. 11).

Figure 11: A snapshot of the MONI.CA application.
Visit http://www.monicapmslivorno.eu. Last access on May 12th, 2016.

The platform is intended to integrate into a unique tool both monitoring and controlling services.

Specific functions are related to monitoring of ordinary and dangerous goods, to convey mobile information to passengers and workers, to monitor road and rail logistics resources, to monitor network infrastructure, to perform risk assessment and crisis management on the landside (including the dry port) and in port waters, to monitor environmental variables for the prompt detection of chemical contamination, and to implement video surveillance (triggered upon the occurrence of specific critical events).

MONI.CA diversifies its services along two dimensions: the reference zone (i.e. port, inner road network, dry port, city access, highway, etc.), and the service type. For the service types MONI.CA is designed to support applications related to safety, security, tracking, navigation, environment, info-mobility and maintenance.

- MONI.CA is designed to manage information relevant for implementing risk avoidance strategies for ordinary work activities and for those in presence of dangerous goods. Data are extracted from sensors (e.g. chemical, radiation), OCR-based equipment, RFIDs and video-surveillance (either via cameras or via drones). Anti-collision systems are expected to play an active role in the field. To the same extent, similar sensors can be used to profile access and use of certain resources to well-defined types of users.
- MONI.CA is expected to track vehicles and freight (stored in cargo trailers and in containers) using OBUs and road/railways resources (interconnected via wireless e.g. via Bluetooth or Zigbee).
- MONI.CA is designed to monitor the position of ships as they navigate in the port area and/or approach the target berths and terminals.
- MONI.CA detects the value of pollutants in real-time (e.g. CO_2, Particulate Matter, SOx, NOx, etc. …); in air and water; it also detects the level of noise in the same areas generating warnings and alerts when certain thresholds are reached.
- MONI.CA is designed to deliver information to final users, e.g. truck drivers and passengers of ferries and cruises; traffic information is aggregated with complementary information coming from the (smart) city information systems, weather information, ship lines schedule and actual time of departure/arrival.
- MONI.CA is designed to support predictive maintenance activities like those intended to fix possible structural problems in berths, railways and docks.
- MONI.CA is also featuring a rendering module in 3D platform for real time port monitoring and control providing an ergonomic interaction with the final user. This does not preclude future M2M interaction getting rid of the human in the loop and boosting on automation.

A pictorial view of the MONI.CA sensing layer is represented in Fig. 12.

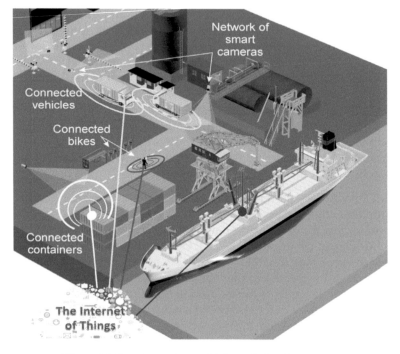

Figure 12: Smart sensing and 3d rendering in seaports.

Apart from representing scalar-valued sensors, MONI.CA also has a data aggregation engine, allowing running high-level business logic and support decision-making processes (both on-line and off-line). MONI.CA will therefore integrate, combine and generate information and statistics as well as being a rendering engine to display geo-referenced results in 3D.

7. Long-term Plans

7.1 Setting up a Service and Control Room

The Livorno Port Authority envisions to integrate and abstract all software applications offered to the users of the port, notably the port community and the port monitoring systems described in Section 6.

In the future port community systems, the workflow to accomplish a logistic and commercial process ideally passes through the brokering, booking and releasing of concrete resources like parking lots, gate entry slots, truck selection, etc.

Functions in PCS can use MONI.CA as a reliable data bank where raw records coming from the sensors and aggregated information built either on-line or via mining processes can be found. The status of port resources together with external inputs will set semaphores in the workflow so that a process is allowed, denied, or postponed waiting for specific events.

7.2 Using Innovation to Set up Corridor-wide Services

Service abstraction is a beneficial Software Engineering approach in a highly dynamic scenarios.

The devised IT architecture will stand the crucial changes that are expected to take place around the Port of Livorno in the near future. Interconnecting the Port of Livorno with the other regional specialized poles (i.e. airports and minor ports) will allow to incrementally build up the so-called Tuscan Logistic Platform, a key objective of the regional economical master plan.

This technical initiative is framed around the political action plan undertaken by the Tuscan executive board which is merging the administrative structures of the two main airports (Pisa and Florence) and the two main ports (Livorno and Piombino) in Tuscany. The Tuscan logistic platform will therefore behave as a unique traffic attractor and source for the import and export respectively (see Fig. 13).

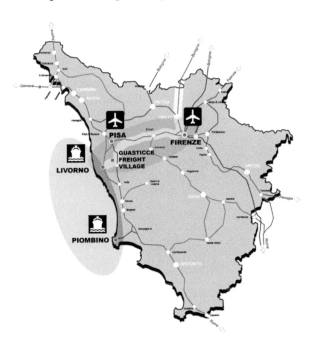

Figure 13: The Tuscan logistic platform.

The Port Community System (extended to the regional scope) will support new logistics model as a turnkey for letting the gateway be faster and low-cost in routing freight and passengers.

Synchro-modality for instance promotes the idea of centering the logistical process on the transfer of goods instead of considering silo-styled shipments combining goods and carriers in a static way.

In a synchro-modal shipment different modes of transport in a network (i.e. those offered by the Tuscan logistical platform) are dynamically combined so that the customer (shipper or forwarder) is offered an integrated solution for his (air/maritime/inland) transport.

Same considerations apply to passengers transport when an effective integrated and inter-modal set of services is offered through the usual ICT channels (i.e. smartphones apps, broadcasts and information publication via the web).

8. Appointed Programs, Partnership and Project Drivers

As mentioned in Section 7, the Port Authority has committed to lead the process of defining a transport and logistics platform at the regional scale. It will orchestrate this process by means of a Service and Control Room hosting the stack of processes on the cloud introduced in Section 3.

To accomplish this task, the Port of Livorno can be regarded as a playground where ICT solutions can be tested and validated.

To this extent CNIT has got approved by the European Telecommunications Standard Institute (ETSI) a proposal to host the forthcoming ITS Plugtests™ campaign* (scheduled in November 2016) at the Port Authority premises. The scenario extends to the local road network including the Control Room of the highways operator.

Figure 14: The Plugtests™ 2016 scenario as from the CNIT proposal. (See http://goo.gl/Pz6JnD. Last access on May 12th, 2016.)

---

* ETSI plugtests reference site: http://www.etsi.org/about/what-we-do/plugtests Last access on May 12th, 2016.

The final target is that of inviting vendors of ITS equipment to converge to a real-world (relevant) example; the benefit of the Port Authority will be that of experiencing the readiness of solutions complying with the standards both for hardware products and software solutions and services.

In the CNIT plans, the Plugtests™ campaign will feature an IoT and M2M showcase in order to demonstrate the relevance of bridging the distributed sensing and the smart mobility domains.

Especially considering M2M that is a declared horizontal Key Enabling Technology, the port opens to the innovation for inter-modal services following the directives towards modal-shift coming from the European Commission and national governments.

Storage and custody of big data will also be taken in consideration especially when the seaport is (inter-modally) connected with other points of interest like airports, stations, city centers, and other seaports along trans-national corridors and the Motorways of Sea.

In this perspective, the Plugtests™ campaign is felt as a pivotal point, a milestone in the roadmap to let the port acquire the innovation needed for a sustainable growth.

9. Conclusions

In this chapter, a possible roadmap for the ICT innovation of the seaports is discussed.

Through the case study of the Port of Livorno, the design and implementation of IoT and ITS applications is discussed in respect of the long-term plans by the Port Authority.

The focus is that of allowing added-value processes through the development of new functions in the Port Community Systems, the popular portals accessed and operated by the Port Communities, and the Port Monitoring System, an application offering a real-time photograph of the port (in terms of sensor readings and processes status).

The Port of Livorno is carrying on an ambitious project of leading the implementation of a multi-level regional transport and logistic platform by leveraging the integration of modes transport as from the political actions undertaken by the Tuscan executive board.

Hosting an international standardization event (that of ETSI Plugtests™ campaign in 2016) is considered a turnkey to shorten the technology transfer process towards the real world realm of the ports.

References

1. The SAFESEANET platform: http://www.emsa.europa.eu/ssn-main.html Last access on May 12th, 2016.

2. EFFORTS (FP6-031486) Effective Operation in Ports.
3. MOS4MOS (2010-EU-21102-S) Monitoring and Operation Services for Motorways of the Sea – Selected as one of the twenty best TEN-T projects 2010.
4. GREENCRANES (2011-EU-92151-S) Green Technologies and Eco-Efficient Alternatives for Cranes and Operations at Port Container Terminals.
5. B2MOS (2012-EU-21020-S) Business to Motorways of the Sea.
6. MONALISA 2.0 (2012-EU-21007-S) Securing the Chain by Intelligence at Sea.
7. MED.I.T.A. (2C-MED12-13) Mediterranean Information Traffic Application.

Index

T - #0442 - 071024 - C208 - 234/156/9 - PB - 9780367782825 - Gloss Lamination